Video Editing Made Easy with DaVinci Resolve 18

Create quick video content for your business, the web, or social media

Lance Phillips

BIRMINGHAM—MUMBAI

Video Editing Made Easy with DaVinci Resolve 18

Group Product Manager: Rohit Rajkumar
Publishing Product Manager: Nitin Nainani
Senior Editor: Keagan Carneiro
Senior Content Development Editor: Debolina Acharyya
Technical Editor: Simran Udasi
Copy Editor: Safis Editing
Project Coordinator: Manthan Patel
Proofreader: Safis Editing
Indexer: Pratik Shirodkar
Production Designer: Alishon Mendonca
Marketing Coordinators: Nivedita Pandey, Namita Velgekar, and Anamika Singh

First published: March 2023

Production reference: 3120925

Published by Packt Publishing Ltd.
Livery Place
35 Livery Street
Birmingham
B3 2PB, UK.

ISBN 978-1-80107-525-1

www.packtpub.com

I would like to dedicate this book to my dentist, who did a wonderful job saving my teeth and, with his wife's permission, also contributed their wedding video for this book.

I would like to dedicate this book to my friends, who supported us when we were temporarily homeless while writing it by giving us a place to stay, as well as their moral support.

Finally, I would like to dedicate this book to my wife, who helped me write this dedication and loves my new smile but, especially, our new house.

– Lance Phillips

Foreword

About a year ago, a colleague recommended Lance Phillips as a DaVinci Resolve-certified trainer to teach a module on our MA editing and post-production course.

From the moment I met Lance, it was clear that we both shared a passion for education and using our personal career experience to give our students the tools and confidence to progress in their chosen fields.

Our MA students have found Lance's teaching methods both engaging and appropriate to their varied abilities and educational backgrounds.

In *Video Editing Made Easy with DaVinci Resolve 18*, Lance has combined his knowledge, acquired through many years of experience across filmmaking disciplines and as an educator, to give a new generation of video-makers the ability to create professional high-quality content.

By using freely available software, *Video Editing Made Easy with DaVinci Resolve 18* delivers a comprehensive yet accessible training tool for the aspiring social media director and editor.

It provides an essential step-by-step guide, from beginner to advanced level, that will enable you to produce high-quality video content with freely available resources.

Stefania Marangoni

Senior lecturer and MA editing and post-production course director at London South Bank University

Contributors

About the author

Lance Phillips is a UK-qualified teacher, actor, color-grading lecturer, Blackmagic Design Training Partner, and Certified DaVinci Resolve Trainer, with 30 years' experience in training and supporting people from diverse backgrounds (including master's degree students, young people, prisoners, people with disabilities, and people from diverse ethnic backgrounds) in their creative careers.

As a Blackmagic Design Training Partner for DaVinci Resolve software, Lance specializes in training media/film professionals at the early stage of their careers or those who are new to using Resolve, which enables them to make their own films and digital experiences to tell their own stories.

In addition to delivering DaVinci Resolve training to film industry professionals, Lance delivers a color-grading module at the master's level for **London South Bank University (LSBU)**. He is also a creative technology researcher for LSBU's Research and Development Hub at Maidstone Studios, where he collaborates with small and medium-sized businesses on research on virtual production and volumetric capture, among other emerging technologies.

Special thanks to my friends who supplied practice media for this book:

Dr. Awais Ali, Dentist and Videographer of filmdental.com, for supplying his wedding video for chapters 7, 8, 10, and 11.

Donovan Parsons, Multi-Skilled Lighting Cameraman of ITN and SET-EST LTD, for supplying the greenscreen footage for chapter 9.

Kathryn de Vries, Blackmagic Design, for her support to me as a trainer to keep my skills and knowledge updated and supplying a studio version of Resolve for the Technical Editor to check Chapter 12.

About the reviewer

Alex Berry has been working in post-production for over 20 years, starting at an independent post-production house and then as a national broadcaster, before founding Quality Control TV, a boutique color-grading studio based in London. He has worked for many leading brands at an international level. As a self-taught editor and colorist, he has always needed to personally seek out information and resources to improve his craft, which is why he's happy to endorse this book, which will help you to do the same. Alex presents a course on color grading in DaVinci Resolve, which you can find on the Domestika learning platform, following on from the excellent foundation provided by this book.

Table of Contents

2

Adding Titles and Motion Graphics   37

3

Polishing the Camera Audio – Getting It in Sync   63

4

Adding Narration, Voice Dubbing, and Subtitles — 89

5

Creating Additional Sound — 131

Part 2: Fixing Audio and Video

6

Working with Archive Footage   157

7

Stabilizing Shaky Footage   171

8

Hiding the Cut – Making Our Edits Invisible   187

Part 3: Advanced Techniques

9

Adding Special Effects   207

10

Split Screens and Picture-in-Picture   233

11

Enhancing Color for Mood or Style   255

12

Studio-Only Techniques   275

Glossary

295

Answers to Questions   303

Index   307

Other Books You May Enjoy   316

Preface

DaVinci Resolve is used to enhance the color of Hollywood films, TV shows, and commercials and does so better than any other video editing software. Version 18 enables you to edit, compose VFX, mix sound, and deliver for different platforms, including social media – all in one piece of software.

This book provides a hands-on approach to using DaVinci Resolve to create and enhance your videos for social media and the web, using a Hollywood-standard video editing suite of tools.

Who this book is for

This book is for emerging creatives, social media influencers, and content makers; anyone with an idea they want to realize and publish online will benefit from this book.

Not only will the nascent content maker who is looking for quick techniques to improve their work benefit from this book but it will also be useful for experienced content makers who want to begin using DaVinci Resolve for its more advanced features. It is a useful and practical training manual for social media marketers, influencers, short film makers, small business owners creating their own content, vloggers, and film/media studies students at schools, colleges, or universities.

What this book covers

Chapter 1, *Getting Started with Resolve – Publishing Your First Cut*, will teach you how to quickly import, organize, review, assemble, and publish video footage on the **Cut** page's **Timeline**.

Chapter 2, *Adding Titles and Motion Graphics*, will help you enhance the project you created in *Chapter 1* by adding a basic title and motion graphics, using **Fusion FX** and transitions on the **Cut** page.

Chapter 3, *Polishing the Camera Audio – Getting It in Sync*, looks at how recording separate audio can improve your video and how Resolve makes it easy to sync separate sounds with your video.

Chapter 4, *Adding Narration, Voice Dubbing, and Subtitles*, will look at making our now-complete video accessible by adding narration, voice dubbing, and subtitles.

Chapter 5, *Creating Additional Sound*, will look at creating a sound library and importing sound effects into **Fairlight** to then use on our **Timeline**.

Chapter 6, *Working with Archive Footage*, will look at using **Fusion FX** to restore old video footage and changing the timing of our dubbed audio to match the timing of the video.

Chapter 7, *Stabilizing Shaky Footage*, teaches you how to stabilize shaky video footage on the **Cut** and **Edit** pages.

Chapter 8, *Hiding the Cut – Making Our Edit Invisible*, will look at continuity editing and how to use cutaways, cut-ins, and split edits on the **Edit** page to hide bad cuts.

Chapter 9, *Adding Special Effects*, will cover how to shoot for a green screen and how to work with it in DaVinci Resolve, using different types of Keyers.

Chapter 10, *Split Screens and Picture-in-Picture*, looks at creating a split screen using composite footage and video resizing, as well as picture-in-picture effects using the Resolve FX Transform **Video Collage** filter on the **Cut** page.

Chapter 11, *Enhancing Color for Mood or Style*, will introduce color theory as used in Hollywood films and how to use color FX tools on the **Cut** page to fix color.

Chapter 12, *Studio-Only Techniques*, will look at the AI Neural Engine tools, only available in the Studio version of DaVinci Resolve, that can speed up your edit process.

Chapter 13, *Glossary*, covers the definitions of all the key technical and film-making terms used in the book.

To get the most out of this book

For this book, you do not need to have any previous technical knowledge, apart from a basic understanding of how to upload content to social media and how to organize files on your own computer.

You will need at least the latest version of DaVinci Resolve 18.1 installed on your computer. The exercises in this book should also work with future releases of DaVinci Resolve.

Software/hardware covered in the book	Operating system requirements
DaVinci Resolve 18.1	Windows, macOS, or Linux

You will need the Studio version of DaVinci Resolve 18 for some of the features explained in this book, particularly Chapters 4, 6, 11, and 12.

We advise you to source videos yourself or download the practice video files from the book's Packt server link (a link is available in the next section).

Feel free to post your video creations on social media and use the hashtag #ResolveEasyEdits to share them with us.

Download the example video files

You can download the example video files for this book from `https://packt.link/B5bqz`. If there's an update to the video files, they will be updated.

We also have code bundles from our rich catalog of books and videos available at `https://github.com/PacktPublishing/`. Check them out!

Download the color images

We also provide a PDF file that has color images of the screenshots and diagrams used in this book. You can download it here: `https://packt.link/NtGgR`.

Conventions used

There are a number of text conventions used throughout this book.

`Code in text`: Indicates code words in text, database table names, folder names, filenames, file extensions, pathnames, dummy URLs, user input, and Twitter handles. Here is an example: "Where it says **Save as**, name your `Bins.DRB` file something useful such as `Video Bins wo Credits`."

Bold: Indicates a new term, an important word, or words that you see on screen. For instance, words in menus or dialog boxes appear in **bold**. Here is an example: Make sure you are in **List View**.

> **Tips or important notes**
> Appear like this.

Get in touch

Feedback from our readers is always welcome.

General feedback: If you have questions about any aspect of this book, email us at `customercare@packtpub.com` and mention the book title in the subject of your message.

Errata: Although we have taken every care to ensure the accuracy of our content, mistakes do happen. If you have found a mistake in this book, we would be grateful if you would report this to us. Please visit `www.packtpub.com/support/errata` and fill in the form.

Piracy: If you come across any illegal copies of our works in any form on the internet, we would be grateful if you would provide us with the location address or website name. Please contact us at `copyright@packt.com` with a link to the material.

If you are interested in becoming an author: If there is a topic that you have expertise in and you are interested in either writing or contributing to a book, please visit authors.packtpub.com.

Share Your Thoughts

Once you've read, we'd love to hear your thoughts! Scan the QR code below to go straight to the Amazon review page for this book and share your feedback.

https://packt.link/r/1801075255

Your review is important to us and the tech community and will help us make sure we're delivering excellent quality content.

Download a free PDF copy of this book

Thanks for purchasing this book!

Do you like to read on the go but are unable to carry your print books everywhere?

Is your eBook purchase not compatible with the device of your choice?

Don't worry, now with every Packt book you get a DRM-free PDF version of that book at no cost.

Read anywhere, any place, on any device. Search, copy, and paste code from your favorite technical books directly into your application.

The perks don't stop there, you can get exclusive access to discounts, newsletters, and great free content in your inbox daily

Follow these simple steps to get the benefits:

1. Scan the QR code or visit the link below

https://packt.link/free-ebook/9781801075251

2. Submit your proof of purchase
3. That's it! We'll send your free PDF and other benefits to your email directly

Part 1:
A Quick Start to DaVinci

The objective of this section is for you to gain an understanding of how to use the **Cut** page in DaVinci Resolve as a tool to quickly create a finished video for social media or the web.

This section comprises the following chapters:

- *Chapter 1, Getting Started with Resolve – Publishing Your First Cut*
- *Chapter 2, Adding Titles and Motion Graphics*
- *Chapter 3, Polishing the Camera Audio – Getting It in Sync*
- *Chapter 4, Adding Narration, Voice Dubbing, and Subtitles*
- *Chapter 5, Creating Additional Sound*

1

Getting Started with Resolve – Publishing Your First Cut

As a video content maker, whether you are creating videos for the web, YouTube, or TikTok, there will come a point where you will need to edit the videos you have made using a dedicated video editing software, rather than struggle with the limitations of a phone app. Dedicated video editing software applications can be quite daunting for a beginner, but once mastered, they can provide a whole host of powerful features to take your videos to the next level.

There are many video editing applications to choose from. This book will cover how to use Davinci Resolve, which is a powerful video editor used by the film and TV industry to edit, color, and add sound and visual effects to films. It is amazing that Resolve, which used to cost $1,000s, is now available for free!

Resolve is an incredibly powerful piece of software. In fact, it is not just one software program, it is a merger of several very powerful software programs that the film industry has been using separately for decades. Each page in Resolve has the power of each one of these software programs. We will cover some of these pages later in the book, but as a beginner, we will be mostly focusing on using the **Cut** page to quickly and simply create your edited video.

The **Cut** page is an incredible tool to quickly put together your first video edit or *cut*. It has many tools available in the rest of DaVinci Resolve but simplified onto one page, to make them quicker and easier to use.

In this chapter, you will learn how to quickly import, organize, review, assemble, and publish video footage on the **Cut** page **Timeline**. Additionally, you will learn how to start a new project and set and save your project settings for future use. You will be shown the interface of Resolve and the **Cut** page, learn how to customize it, and be able to save your preferred settings so that you can have a user workspace that suits you. You will then import media that you want to edit together (i.e., audio, video, photos, and computer graphics) and use the unique features of the **Cut** page such as **Source Tape** and **Smart Insert** to review and add footage to the **Timeline** to quickly create a basic edited video. Finally, you will export and publish a video directly to YouTube from within the **Cut** page of DaVinci Resolve rather than having to export it first and then upload it later.

All these tools combined will make editing and publishing short-form films and videos much quicker, saving you time and freeing you up to make more content.

In this chapter, we're going to cover the following main topics:

- Creating your first project in DaVinci Resolve

- Laying out your digital workspace how you want it

- Getting your media files in and organized

- Reviewing your shots and cutting it all together

- Publishing your video to social media

> **Fun fact**
>
> The editing term **Cut** is named after the physical cutting process used in the early years of editing film, where a razor blade was used to cut out unwanted footage before it was all taped back together for the final edit. Today, it is a lot safer just to use DaVinci Resolve to *cut* your films digitally!

Technical requirements

You will need to have installed a copy of DaVinci Resolve version 18. Some exercises may work in older versions of Resolve after version 16. However, to get the most from this book, it is best to have the latest version of DaVinci Resolve. You can download the latest free version of DaVinci Resolve from the Blackmagic Design website here: `https://www.blackmagicdesign.com/uk/products/davinciresolve/`

Also, download the **DaVinci Resolve Bins (DRB)** file here: `https://packt.link/B5bqz`

All exercises will also work with the paid studio version of DaVinci Resolve. The technical requirements for your computer to run DaVinci Resolve can be found on the Blackmagic Design website here: `https://www.blackmagicdesign.com/support/readme/e8b376651a8d4f1fb7bb18167325fb7f`

Creating your first project in DaVinci Resolve!

When opening Resolve for the first time, it can be quite confusing where to start. You will learn how to create a new project and where to set and save project settings in DaVinci Resolve so that you can load them up quickly for future projects.

Opening Resolve for the first time

After installing Resolve, the first time you open it, you will be greeted with the welcome screen (*Figure 1.1*) that gives you an overview of the latest features. This can be found in the **Help** menu if you need to be reminded: **Help > Welcome to DaVinci Resolve**:

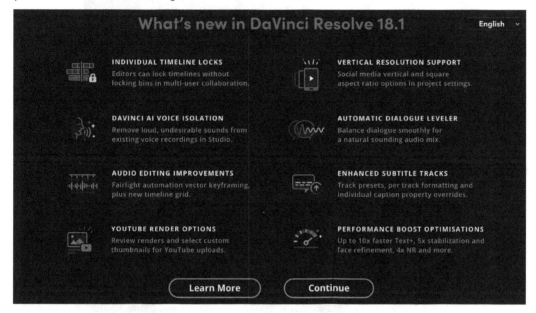

Figure 1.1: Welcome screen

Next, Resolve will ask where you want to save the certain files it needs to be able to work. You can always change these locations later, but first, let us look at what they are so that you can better understand where to tell Resolve to store them.

Cache files

Cache files are media files that Resolve temporarily stores for easy access to help Resolve access them quicker while you are editing, which also helps for the faster playback of your media on your **Timeline**. It is best to store these cache files on your internal hard drive on your computer, or if it's faster, an external **Solid State Drive** (**SSD**) connected to your computer. If you're using an external drive for your cache files, remember to plug it in before you start Resolve; otherwise, Resolve will not know where to store your cache files and present the following message (*Figure 1.2*):

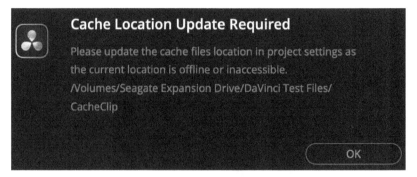

Figure 1.2: Cache files warning message

Stills

Resolve also needs to know where to save your stills. Otherwise, you will get the following message (*Figure 1.3*) if Resolve can not find the drive your stills folder is linked to:

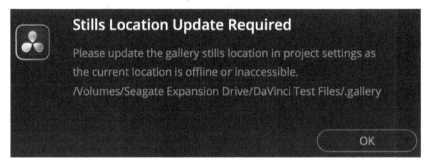

Figure 1.3: Stills warning message

This is not, as it sounds, where you store still photos you use in your edit. It is where Resolve stores still images of frames you select from your **Timeline**. It's a bit like a screenshot or photo of a moment in time of your edit.

This can be useful if you want to share an image from your **Timeline** as a thumbnail for your social media posts. You can store these stills on a hard drive anywhere. It doesn't need to be a fast drive – save that for your cache files.

> **Top tip**
>
> If you change your mind, you can change the cache and stills locations later. It is good practice to store the cache and stills files directly on your computer as Resolve will not lose the link to them. Unless, of course, you move the folders when tidying up your computer. In this case, you can relink them to the moved folder or create a new folder to link to if you have deleted it.

Creating your first project

Now that we have told Resolve where to store the files it needs to create in order to work fluidly, we can create your first project. This is done in **Project Manager** (*Figure 1.4*):

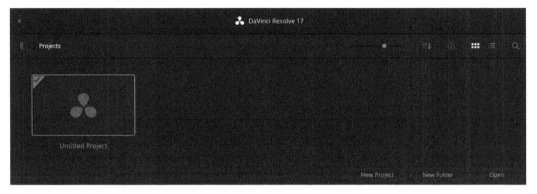

Figure 1.4: Project Manager

There are two ways to create a new project:

- Click on the **New Project** button in the bottom-right corner of the **Project Manager** window
- Alternatively, open the **Untitled Project** template

Which you choose is up to you.

The only difference is that **New Project** will ask you for a project name before creating it, whereas, with **Untitled Project**, you can create a blank project without having to name it.

The **Untitled Project** option is good for creating a temporary project to play around with the features in Resolve without needing to save a project. Of course, if you change your mind, you can always save this project and Resolve will ask you to name it first so that it can save it.

Now that you have created your first project, let's look at the visual layout or **User Interface** (**UI**) of Resolve and how you can customize it to suit how you work.

Customizing your digital workspace

When first using Resolve, it can be a little confusing, as there are so many pages all with a myriad of functions, which to a beginner can be understandably overwhelming. It's like learning to fly for the first time using a Jumbo Jet! Thankfully, we can simplify the workspace by hiding many of the pages so that our interface is more appropriate for a beginner, much more like learning to fly in a small propellor plane. It is still daunting but much more achievable!

However, first, let us look at how Resolve is unique in using pages in the first place.

Understanding the Resolve pages

Blackmagic Design, the owner of DaVinci Resolve, has created a separate page for each of these different software programs laid out in a row at the bottom of the software interface (*Figure 1.5*). The benefit of this is that you can do advanced sound, color, and visual effects without having to export your film edit out into another application. Resolve is the only editing software that has this function. Other editing programs rely on plugins or exporting the edit into another application to do more advanced work and then reimporting it back in to do further edits. Of course, if you prefer to work this way, then Resolve will let you do this too. However, this is a more advanced and time-consuming technique and is beyond the scope of this book.

You might have noticed that the pages are laid out in the order of a traditional filmmaking workflow. However, the beauty of Resolve is that you can go back and forth between the pages and work in any order that you want.

In this chapter and for most of the book, we will be using the **Cut** page, which is the default page that Resolve will take you to when you first open a new project.

The **Cut** page is used for creating a quick edit and is great for quickly creating and publishing social media content or **rough cut** for a feature film:

Figure 1.5: The Cut page

Now, let us simplify our workspace so that we can focus mainly on the **Cut** page without the distractions of the other pages.

Simplifying the workspace

Having so much software functionality can be great when you need to use all of the powerful features of Resolve for your feature film. However, if you just want to use one or a few of the pages, Resolve has a handy feature where you can temporarily hide the page you don't want. Let's do that now. For now, we are going to hide the **Media**, **Edit**, **Color**, **Fusion**, **Fairlight**, and **Deliver** pages:

1. Select the **Workspace** menu in the top-right corner of Resolve (the **Menu** bar).
2. Select **Show Page**.
3. Click on the pages you want to hide – in this case, the **Media**, **Edit**, **Color**, **Fusion**, **Fairlight**, and **Deliver** pages.
4. There will now be a tick next to the pages Resolve will show and no tick next to the pages Resolve will hide. By default, all pages are preselected.

Now that your workspace is a lot simpler, we can save this new look as a preset, which you can quickly load again when you want to use it in the future:

1. In the **Menu** bar, select **Workspace**.
2. Select **Layout Presets** > **Save Layout as Preset…**.
3. In the **Enter Preset Name** textbox, name your preset `Basic Workspace`.
4. Click on the **OK** button or press the *Return* key on your keyboard.

Well done! You have now created your first workspace preset, which you can quickly recall later.

> **Top tip**
> You can save your layout presets on a USB stick (or even in cloud storage such as Google Drive or DropBox) so that when you are working on someone else's version of Resolve, you can load your workspace without affecting the way *they* like to lay it out.

Now that you have simplified your workspace, let's start importing your media.

Getting your media files in and organized

The **Media** page has some advanced functions for importing media that big-budget feature films use. However, it is quick and easy to import media directly into the **Cut** page. In fact, every page has the space to quickly import media – it's called the **Media Pool**. The **Media Pool** is located in the upper-left corner of every page (*Figure 1.6*), which makes it easy to locate.

Import Media

In the upper-left corner, underneath the **Media Pool** button of the **Cut** page, there are two icons that allow you to quickly import media into Resolve:

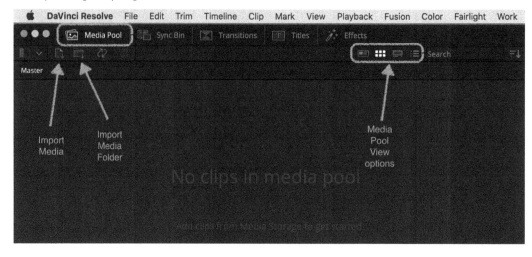

Figure 1.6: Media Pool

The first button, **Import Media** (*Figure 1.6*), imports media files:

1. Left-mouse click on the **Import Media** button.

2. In your computer's **File Manager** window that pops up, navigate to where you have your media files stored.

3. Select the files you want to import.

4. Click on the **Open** button.

5. If your imported footage (clip) does not have the same **frame rate** as your project's frame rate, you will get a pop-up message (*Figure 1.7*) asking you to change the project frame rate to match the footage you are importing:

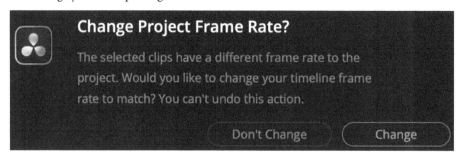

Figure 1.7: The Project Frame Rate pop-up message

6. Go ahead and click on **Change** to change the project frame rate to match the frame rate of your imported clips. We will cover project frame rates in more detail later.

You should notice that your **Media Pool** is now filled up with the media you just imported.

Now, let us look at the different ways we can view our clips in the **Media Pool** so that we can check we have the right footage before moving it to the **Timeline**.

Media Pool – changing your views

By default, your imported media are shown as thumbnails, but you can change this at any time. Let's do this now:

1. Click on the **Media Pool View** options (*Figure 1.6*) in the top-right corner of the **Media Pool**.

2. Select **List View** (the fourth button from the left).

The thumbnails disappear but you can now see a more detailed list of the file properties of the media you imported.

The most common file property is the name of the file, "filename." The problem with most cameras or audio recorders is that they name the files they create with a semi-random name such as SA478937. mov. This probably means a lot to the camera but means nothing to us.

Of course, the temptation is to change the filenames on our computer to make the file content easier to understand. The problem with this approach is that each time you change the filename on your computer (after you have imported it into Resolve), Resolve can no longer find it because it is looking for the previous filename.

A better approach is to use a feature in Resolve called **Clip Name**, which means you can change the name of the media in your bins as often as you like while retaining the original filename on your computer:

1. Make sure you are in **List View**.

2. Select **Clip Name**.

3. Type in the **Clip Name** field to change the clip name to something memorable.

4. Press the *Return* key on your keyboard or click outside the text field.

You have now changed your clip names to easily identify what they are.

Top tip

If you want to see a combination of the **Thumbnail View** and the **List View**, with a list of the most common properties of the media file, select the **Metadata View**.

Now, let's preview some of the files that you have imported to see whether they were the ones you wanted to import:

1. Select **Thumbnail View** again (the second button).
2. Select a clip to load it into the **Viewer**.
3. Hover your mouse over the thumbnail.
4. Move your mouse cursor over the thumbnail.

As you move your mouse cursor over the thumbnail, you will notice the media will play back in the **Viewer** window.

This is a quick way to review your footage before committing it to the **Timeline**. There is an even quicker way if you have a lot of media to review, which we will show you later.

A few words about bins

By default, Resolve has a **Master bin** in the **Media Pool** that contains all the media (video, audio, and graphics) you have imported into your video project.

You might be wondering why it is called the **Master bin**, as like in real life, a bin on a computer is usually where we delete files.

Bin is a word that professional film editors use for a folder. It comes from the early days of celluloid film editing where strips of film would be pegged up over a bin, waiting for the film editor to splice them into the main edit (*Figure 1.8*):

Figure 1.8: Old 1925 photo of film over a bin (Source WikiCommons)

Creating a new bin

You can create extra bins to keep your media organized, for example, one bin for separate audio such as music and sound effects, and another one for video. Let's do that now:

1. Right-mouse click anywhere on a blank space inside the **Master bin**.

2. Select **New Bin**.

3. A new bin will be created in the **Master bin**.

4. Name the bin Video. There are two ways to do this:

    *   Right-click on the bin and select **Rename Bin**

    *   Type directly in the **Bin Name** field under the **Bin** icon

5. Create another bin using the preceding steps and name it **Audio**.

> **Top tip**
>
> You can rename any bin apart from the **Master bin** by right-clicking on the bin and selecting **Rename Bin**. Right-clicking on objects such as bins, media, **Timeline**, and Viewers will usually reveal a hidden options menu. Try it and see what hidden options you can discover!

Now that we know what a bin is, and we have created our own, let's look at the second button to import media with, called **Import Media Folder**.

Import Media Folder

This imports media but also imports the folders that are on your computer and converts them into bins:

1. Click on the **Import Media Folder** (*Figure 1.6*) button.

2. In the pop-up window, navigate to where your media files are stored.

3. Select the folder and files you want to import.

4. Click on the **Open** button.

You will now notice that nested inside the **Master bin** are other bins with the same name as the folders that held the media on your computer.

It is good practice to put all your media into appropriately named folders on your computer and import them using this second option rather than the time-consuming approach of creating your bins from scratch every time you edit. The other way to save time is to import the bins that have already been created in another project.

Importing bins

Let's import some of the bins that I have created for you:

1. From the menu bar, select **File** > **Import** > **Bin**….

2. In the pop-up window, choose where on your computer you want to import your bins from.

3. You are looking for the `Lances_Bins.DRB` file you downloaded onto your computer.

4. Select **Open**.

You will now see a series of bins in your **Master bin** called **Video**, **Audio**, **Credits**, and **Graphics**. Each one has a blank media file in it. This is because Resolve needs content in the bins to be able to successfully export or import them. You can delete this temporary media by selecting it and then pressing the *Delete* key on your keyboard.

Deleting bins

You most likely are not going to need the **Credits** bin, so let's delete it:

1. Right-click on the **Credits** bin.

2. Select **Remove Bin**….

3. The bin should now be removed.

Selecting the bin with your mouse and pressing the *Delete* key on your keyboard will also delete the bin.

Exporting bins

If you have a favorite layout for your bins, as of version 17, you can now export your bins structure as a .DRB file (which stands for DaVinci Resolve Bins) and store this on your USB with your UI presets. Let's export your new bin structure without the **Credits** bin:

1. Click on the **Bin List** button (or the drop-down arrow next to it) to reveal a list of bins (*Figure 1.9*):

Figure 1.9: Bin List

2. Right-click on the name of the bin you want to export.

3. In the pop-up window, choose where on your computer you want to export your bins.

4.   Where it says **Save as**, name your `bins` `.DRB` file something useful, such as `Video Bins wo Credits`.

5.   Select **Save**.

Now you can create, delete, rename, import, and export bins to better organize and keep track of all your project's media. Now, let's start reviewing and editing our media.

Reviewing your shots and cutting it all together

The **Cut** page has some fantastic features that allow you to quickly review and edit video footage together. We will go through each step one at a time, but in reality, once you have mastered these tools, each step will only take seconds to execute, making your edit much speedier.

Reviewing your footage

You might recall how we used the mouse to scrub through each video clip's content quickly. However, this can be time-consuming if we must do this for every clip. I am glad to say that Resolve has a far more efficient way to review all your video footage at once, using **Source Tape** mode.

Source Tape

Source Tape (*Figure 1.10*) is a way to watch all your "Source" video footage together at the same time as if playing it as one continuous video tape, hence the name **Source Tape**.

Let's see how it works.

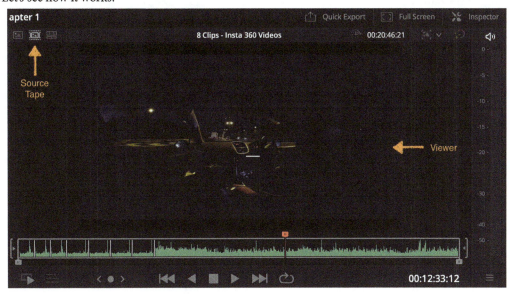

Figure 1.10: Source Tape Viewer mode

1. Select the **Cut** page.

2. In the top-left corner of the media viewer, select the **Source Tape** button (*Figure 1.10*).

3. The **Viewer** (*Figure 1.10*) will now populate with all the media in the selected bin.

4. Scrub through the mini transport **Timeline** under the **Viewer** to review all your footage quickly without having to load each individual clip into the **Viewer**:

    *   The spacebar will play back footage at normal speed.

    *   Pressing the spacebar again will stop the video playback.

    *   Pressing the *L* key on your keyboard will also play through the footage at normal speed.

    *   Pressing the *L* key twice will fast forward twice as fast. Three presses will be 4x as fast, four presses will be 8x as fast, five presses will be 16x as fast, and six presses will be 32x as fast.

    *   Pressing the *J* key on your keyboard will play the footage backward.

    *   Pressing the *J* key twice will fast reverse twice as fast. Three presses will be 4x as fast, four presses will be 8x as fast, and so on.

Top tip

Although I have focused on using keyboard keys to play back our footage (as it is quicker than clicking back and forth with a mouse), you can use a mouse to click on each of the video playback controls (also called transport controls, *Figure 1.10*) underneath the **Viewer**. These are the same symbols you will find on your CD/MP3 or DVD player, if you have one, and they work in the same way.

You will have noticed that, in **Source Tape** mode, the bin icons have disappeared and been replaced with dates instead. These are the dates when the video footage was created. You can change this view to show the bin names by selecting the **Sort** button. Let's do this now.

5. Click on the **Sort** button (*Figure 1.11*):

Figure 1.11: The Sort button

6. From the drop-down list, select **Bin**.

You can isolate the playback of the footage to just the contents of one bin rather than everything in the **Master bin**, perhaps to review footage of a bin you have just imported:

1. Select a bin from the **Master bin** using the **Bin List** icon (*Figure 1.9*) in the top-left corner of the **Media Pool**.

2. Select **Source Tape** mode again (if it isn't already selected).

3. Play back and review the footage by pressing the spacebar or the *J* and *L* keys.

You will have noticed several things when playing back footage in **Source Tape** mode:

- What gets played back in **Source Tape** is dependent upon what bin you have selected

- The vertical lines in the **Viewer Transport Timeline** (*Figure 1.10*) represent the beginning and end of clips

- As each clip is played back, it is highlighted (with an orange border) in the **Media Pool**

Now you can see how useful **Source Tape** is for reviewing all the footage in your master or selected bins quickly rather than having to select each clip individually.

However, there is an even quicker way within **Source Tape** that you can use to review your footage, called **Fast Review**.

Fast Review

Fast Review uses simple **Artificial Intelligence** (**A.I.**) to adjust the playback speed of the clip based on the clip length. So, long clips are played back faster, while short clips are played back slower so that you don't miss them. Let's try it:

1. Make sure **Source Tape** is selected.

2. Select **Fast Review** (*Figure 1.12*) from the far left of the playback controls under the **Viewer**:

Figure 1.12: Fast Review mode

Fast Review gives you a quick and easy way to review your video footage to help you make decisions on what to include in your edit.

Editing your footage together

Before you edit your footage, let's describe what a **Timeline** is and how it uniquely works on Resolve's **Cut** page.

The Cut page timelines

All editing software has what is called a **Timeline**. This is a long horizontal strip/line onto which footage (frequently referred to as clips) is dropped, with the time or duration of the edited video displayed along the top of the **Timeline**.

> **Timelines**
>
> Use of the word Timeline is applied to both the horizontal strip where clips are edited (on the interface) and the name of the file in the **Media Pool** that contains all your editing decisions.
>
> In this book, to differentiate between the two, we use Timeline for the Timeline interface and timeline for the timeline file in the **Media Pool** or when talking about timelines in general.

Timecode

This time is displayed in hours, minutes, seconds, and frames and is called a timecode. It is displayed as HH:MM:SS:FF (*Figure 1.13*):

Figure 1.13: The Timeline timecode

Upper timeline

What is unique to the **Cut** page is that it has two Timelines (*Figure 1.14*). The topmost **Upper Timeline** is the overview of your entire edit, which displays the whole duration of your entire video regardless of how long it is and how many clips there are. This is great for quickly navigating to any part of your **Timeline** and seeing how all your clips are edited together:

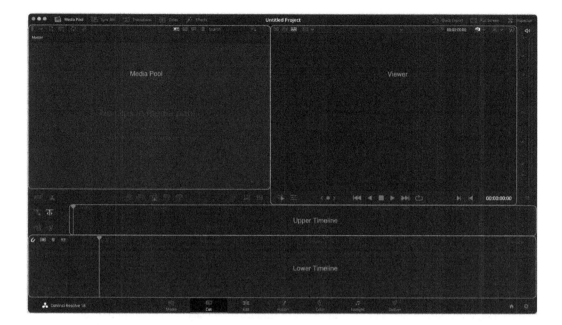

Figure 1.14: Cut page Timelines

Lower Timeline

The second **Timeline** (**Lower Timeline**), just below the **Upper Timeline**, is a more traditional editing **Timeline** that shows you a zoomed-in view of your **Upper Timeline**, which is a lot easier to see and make your edits with.

Key concept – Timelines, tracks, and shots

In a **Timeline**, each layer of the video is called a track, and each track is made up of a series of clips representing different camera shots or angles. So, the hierarchy of content goes like this:

Project > **Timeline** > **Track** > **Clips**.

So, a project can have several different Timelines, each **Timeline** can have several different tracks, and each track can contain several different clips.

Track 1 versus Track 2

On the **Cut** page, Track 1 (the bottom track) operates differently from Track 2 and above (that is, all the tracks above/on top of Track 1).

Track 1 will automatically shuffle footage (clips) along the **Timeline** to make room for the new clip – this is called **Rippling**. This is great for quickly adding clips to the **Timeline** to get your first overall edit done, also known as a **Rough Cut**. Whereas any clip added to Track 2 will overwrite any clips already on Track 2. Track 2 is useful for adding additional clips without affecting the duration of your first edit, such as adding cutaways or titles to be shown over existing clips.

> **Key concept – cutaway**
>
> We will cover cutaways in more detail in *Chapter 8*. However, as a brief overview, a cutaway is footage that is added to the edit that illustrates already existing footage on the **Timeline**.
>
> An example of this is when a person is being interviewed and they refer to a product they are promoting – this is a good place to add a shot of the product on a track on top of the original interview footage where we continue to hear the interviewee describe the product.
>
> The effect of this would be that the viewer would see the interviewee talk about the product, then see the camera "cut away" to a shot of the product while we still hear the interviewer talking about it. This gives the illusion to the viewer that these shots are happening in real time, as if they are watching a live multi-camera broadcast switching between cameras.
>
> Of course, none of this has to be shot in real time and can all be accomplished in the edit.

Let us see how Track 1 rippling works in practice.

Adding footage to the Timeline – using Track 1

Using the skills you have just learned, use **Source Tape** and **Fast Review** to identify five shots you would like to use for your video:

1. Using **Fast Review**, find and identify the followings shots:

    - The master shot (a wide shot)

    - Two mid shots of your main subject

    - Two close-ups (either a reaction shot or a cutaway shot)

2. Select your master shot in the **Viewer** window.

3. Click and drag the master shot onto either **Timeline**.

Now, you have a master shot on the **Timeline**, which you can use to edit your other shots. All clips can be added to the **Timeline** by either clicking and dragging directly from the media bin or from the **Viewer** onto the **Timeline**.

Key concept – shot types

When making a video, we use different framing conventions called shot types, which add visual interest for the viewer by changing how much of our subject is in the frame. The closer the shot is to the subject (which is usually a person), the more it has an emotional impact on the audience.

The three main shot types are as follows:

- **Wide Shot (WS)**: This shot contains the whole environment within which our subject is positioned and how they interact with it. For example, it could be a person standing in a room, where we see all of their body and their relationship to their environment – that is, their home or workplace. This shot is too wide to invoke much emotion.

- **Mid/Medium Shot (MS)**: The bottom of the frame starts at the waist of our subject and the top of our frame ends just above the head of our subject. That is good for formal interviews.

- **Close Up (CU)**: This is a small detail that is magnified, such as framing just the head and the top of the shoulders of the subject. This is good for showing intense emotion. It could also be a tight frame of our subject's hands demonstrating a product.

There are many more shot types, but these are the main ones to use when you are just starting out. If you change your mind, you can delete any clip from the **Timeline** by selecting it with the mouse and pressing the *Delete* key on your keyboard. Don't worry – doing this will not delete your original footage as it is still available in the media bin.

Key concept

Editing in Resolve is "non-destructive," which means that you can delete clips from the **Timeline** and even the media bin and still be able to reimport the same media back in.

When you edit, you are not cutting your original footage but a live preview of it. Resolve stores the editing decisions you make (in a database) and then applies them to your final film when you export it.

The timeline playhead

On the **Timeline**, you will notice a vertical orange line – this is called the playhead (*Figure 1.14*). The **Viewer** will show a picture of the frame that is under the playhead on the **Timeline**. The **Timecode** display in the bottom-right corner of the **Viewer** window also shows the timecode for the current frame under the playhead.

You can move the **Timeline** playhead just like the playhead in the **Viewer** using the spacebar or the *J* and *L* keys.

You can also use the playhead position to tell Resolve where you want to make a cut. Let's remove the usual dead space at the start of your master shot:

1. Position your playhead just before the main action starts in your story.

2. If there is dialogue, you can use the audio waveforms at the bottom of the clip as a guide.

3. Select **Trim Start** from the **Trim** menu (this is called **Trim to Playhead** in earlier versions of Resolve).

Now your story starts by jumping straight into the action.

Usually, your master shot will be far too long to use by itself, so we will need to break it up to bring visual interest to the viewer.

Find a part of your master shot where your subject has finished their main thought or action. We are going to cut it here, and add a different angle to add visual interest:

1. Move the playhead to where you want to make a cut.

2. Click on the **Split Clips** button (*Figure 1.15*) just below the **Media Pool**:

Figure 1.15: The edit buttons

You will notice that the master shot has now been split into two. The dotted vertical line shows that the footage is continuous and has no frames missing – this is called a **through edit.**

Setting in and out points

When editing, we don't have to add a whole clip to the **Timeline**. We can trim the footage down to be shorter before we add it to the **Timeline** by adding **In and Out Points**. In and out points dictate the start (In) and end (Out) of the footage we are adding to the **Timeline**.

Now, let's find an alternative clip to insert between the two clips we already have on the **Timeline** and trim out any excess footage from our new clip, using in and out points, before adding it to the **Timeline**:

1. In the **Media Pool**, select a mid-shot clip that matches or enhances the action of the master shot.

2. Play the clip in the **Viewer** until just before any action or speech. Then, press the *I* key on the keyboard to add an **In** point for the clip to start at.

3. Play the clip forward until just after the last words are said or an action is performed. Then, press the *O* key on the keyboard to an **Out** point for the clip to end at.

You will now notice an in and out point marker added to the thumbnail of the clip in the **Media Pool**, as well as underneath the **Viewer**. Now, when we put this clip on the **Timeline**, it will only take the footage in between the in and out points we have marked up on the clip. Now, let us insert this marked-up clip onto the **Timeline** using a feature called **Smart Insert**.

Smart Insert

Traditionally, when we add a clip to the **Timeline**, it will place it where the **Timeline** playhead is. There is a new feature called **Smart Insert** in the **Cut** page that predicts where we want the new clip to go based on how near the playhead is to a pre-existing cut. Let's use **Smart Insert** to add our marked-up clip to the **Timeline**:

1. Make sure the playhead is near the cut that we made earlier; it doesn't need to be precisely placed.

   You will notice some arrows bouncing above the cut point. These are *Smart Indicators*, that is, they are a visual cue to show you where the clip will be inserted when you click on the **Smart Insert** button.

2. Click on the **Smart Insert** button (the first edit button; see *Figure 1.15*) from the upper-left corner of the **Timeline**.

The clip you have chosen has now been trimmed and added to the **Timeline** at the point where you previously made a cut. You should also notice that the surrounding clips have not been overwritten, but they have been moved further up the **Timeline** to make room for the new clip – this is called *Rippling*

Append to End

Now, let's look at the next edit button (see *Figure 1.15*) in the editing toolbar underneath the **Media Pool**, **Append**.

The **Append** button places a new clip at the end of our **Timeline** regardless of where the playhead is placed. This is great for credits or adding footage quickly to the **Timeline** in chronological order without needing to position the playhead first:

1. In the **Media Pool**, find a new clip that you want to place at the end of the **Timeline**, such as the final shot of the product you are promoting or the final scene of your short film.

2. Click on the **Append** button.

The new clip has now been placed on the end of the **Timeline** regardless of the position of your playhead.

Ripple Overwrite

If you want to replace a clip on the **Timeline** with another clip in the **Media Pool**, then you can use **Ripple Overwrite** (the third edit button; see *Figure 1.15*). This is great if you want to change your mind about the content of your video. When you use **Ripple Overwrite**, if the new clip is too short, it will ripple the rest of the clips around the new clip to close any resulting gaps. If the new clip is too long, it will ripple the existing clips along the **Timeline** to make room for the new clip.

Let's see this in action:

1. Select the clip in the **Timeline** that you want to replace. You can simply place your playhead on the clip to do this.
2. Select the new clip in the **Media Pool**.
3. Click on the **Ripple Overwrite** button.

You should have noticed that the new clip has replaced the old clip on the **Timeline**, and the **Timeline** has moved the clips to either side (rippled) of the new clip, reducing or increasing the length of the **Timeline**, depending on the duration difference between the two clips.

Adding footage to the Timeline – using Track 2

There are no problems with performing your whole edit by only using the **Smart Insert**, **Append**, or **Ripple Overwrite** buttons. However, there will be times when you will need more advanced functions to solve a particular problem; this is where the following edit tools come in.

So far, all of the edits we have been making have been to Track 1 in the **Timeline**. Whatever edit you make to Track 1 of the **Timeline**, the clips will always ripple to make room for the new edit. If you want to add clips to the **Timeline** without the **Timeline** rippling to make room for the addition, you can add the clips to Track 2, which will not ripple the **Timeline**, but instead overlays a clip (on Track 2) on top of existing clips (on Track 1).

Close Up

The next **Edit** tool, **Close Up** (the fourth edit button; see *Figure 1.15*), uses Track 2 to add a close-up of the existing footage on Track 1. It does this by taking a zoomed-in view of the clip on your **Timeline** where your playhead is and adding a 5-second duration clip to Track 2. If the duration between your playhead and the end of the existing clip is shorter than 5 seconds, it will end the new clip at the end of your existing clip.

It is important to note that this is digitally zoomed, so there might be some loss of quality unless you use a clip that has a higher resolution than the **Timeline**, such as a 4K clip for a 1080p **Timeline**.

This tool does not have to be used only for close-ups; for example, it can also convert a wide shot into a medium shot.

Let's take the medium shot that you placed on the **Timeline** earlier and convert some of it into a close-up:

1. Place your playhead on the **Timeline** over the medium shot clip where you want the close-up to start.
2. Click on the **Close Up** tool button underneath the **Media Pool**.

You will notice a new close-up shot placed on Track 2 above the existing shot on Track 1.

This is a useful tool to use in an emergency if you're editing footage where no close-ups were shot and there is no time for a re-shoot.

Place on Top

Now we are familiar with how to place shots on Track 2 so that the **Timeline** does not become rippled, let us look at another editing tool that uses Track 2 to add alternative footage.

The **Place on Top** edit button (the fifth button; see *Figure 1.15*) adds a new clip to Track 2 above the existing clips on Track 1. This is good when adding cutaways or inserting shots to the main edit without affecting the duration of the video on the **Timeline**.

By default, the **Place on Top** tool places the new clip onto Track 2 starting at your playhead position on the **Timeline**.

Let's see this in practice. We will now add one of our cutaway shots to the **Timeline** to illustrate the main dialogue in the video:

1. Place your playhead in the **Timeline** where you want your new clip (e.g., a cutaway shot) to begin.
2. Select your new clip in the **Media Pool**.
3. Click on the **Place on Top** button.

You should now have a new clip on Track 2 of your **Timeline** starting where you had your playhead placed.

Once you have completed your main edit and set the final duration of your video, you can use this tool to add extra shots without affecting the length of your video.

Source Overwrite

The final edit tool underneath the **Media Pool** is the **Source Overwrite** button. This button only works if you have more than one clip with overlapping timecodes. If you do not have footage with timecodes, then this button will do nothing. As this is an editing book for beginners, we will ignore this button.

Feel free to practice using these editing tools until you have mastered them by adding and swapping the footage on your **Timeline** until you are happy with your edit.

In the next section, we will look at how to publish your edited video to social media such as YouTube or Vimeo.

Let's publish!

Let's see how easy it is to publish your edited video directly from Resolve onto social media platforms such as YouTube or Vimeo.

First, we will need to set up Resolve with our social media accounts to enable it to publish directly through them. You will only need to do this once, but it can save a lot of time compared to exporting your video to your computer and then uploading and publishing it to social media separately.

Setting up your social media accounts in Resolve

Resolve has the ability to publish videos directly to your YouTube, Vimeo, and Twitter accounts. First, you need to sign in to your social media accounts in the **Quick Export** menu:

1. Click on the **Quick Export** button in the upper-right corner of the **Viewer** window.

2. From the **Quick Export** screen, select one of the following:

   - YouTube

   - Vimeo

   - Twitter

3. Click on the **Manage Account** button. This will take you to the **Internet Accounts** page.

4. Click on the **Sign In** button for one of the following:

   - **YouTube**

   - **Vimeo**

   - **Twitter**

5. Sign in to your social media account using your usual login details and authorize Resolve to have access to your account.

6. Click on the **Save** button in the bottom-right corner of the **Internet Accounts** page.

7. Repeat *steps 1 to 6* for your other social media accounts.

> Top tip
>
> If you want to access the **Internet Accounts** page again to sign Resolve out of your accounts, you can access it from the DaVinci Resolve menu: **Preferences** > **System** > **Internet Accounts**.

You now have your social media accounts set up and ready to quickly publish your videos to social media.

Publishing to social media within Resolve

Publishing directly to social media is as easy as selecting some options on the **Quick Export** page.

Publishing directly to YouTube

Let's look at the YouTube options first:

1. Click on **Quick Export**.
2. Select **YouTube**.
3. Check the **Upload Directly to YouTube** box.
4. Select the YouTube **Privacy** option from the drop-down list:

    - **Private**
    - **Public**
    - **Unlisted**

5. Select the playlist options from the drop-down list:

    Film & Animation, Auto & Vehicles, Music, Pets &Animals, Sports, Travel & Events, Gaming, People & Blogs, Comedy, Entertainment, News & Politics, Howto & Style, Education, Science & Technology, or **Nonprofits & Activism**

6. Type the title of your video into the **Title** textbox.
7. Type the description of your video into the **Description** textbox.
8. Click on the **Export** button in the bottom-right corner of the **Quick Export** page.
9. Choose a file location for your exported video.
10. Click on the **Save** button.

Resolve has now saved a copy of your exported video to the location you specified on your computer and uploaded and published a copy of the video directly to YouTube using the settings you specified on the **Quick Export** page.

Publishing directly to Vimeo

Now, let's look at the Vimeo options:

1. Click on **Quick Export**.
2. Select **Vimeo**.
3. Check the **Upload Directly to Vimeo** box.

4. Select the Vimeo **Privacy** option from the drop-down list:

    • **Anyone**

    • **Only Me**

    • **People I follow**

    • **People with the password**

5. Type the title of your video into the **Title** textbox.

6. Type the description of your video into the **Description** textbox.

7. Click on the **Export** button in the bottom-right corner of the **Quick Export** page.

8. Choose a file location for your exported video.

9. Click on the **Save** button.

Again, Resolve has saved a copy of your exported video to the location you specified on your computer and uploaded and published a copy of the video directly to Vimeo using the settings you specified on the **Quick Export** page.

Publishing directly to Twitter

Finally, let's look at the Twitter export options:

1. Click on **Quick Export**.

2. Select **Twitter**.

3. Check the **Upload Directly to Twitter** box.

4. Type the description of your video into the **Description** textbox.

5. Click on the **Export** button in the bottom-right corner of the **Quick Export** page.

6. Choose a file location for your exported video.

7. Click on the **Save** button.

Yet again, Resolve has saved a copy of your exported video to the location you specified on your computer and uploaded and published a copy of the video directly to Twitter using the description you specified on the **Quick Export** page.

> Top tip
>
> On the **Quick Export** page, Resolve uses common video settings for social media accounts. However, you can change these settings in the **Timeline Resolution** drop-down list (*Figure 1.16*).
>
> Once you change the timeline resolution, you will notice that the export file settings on the **Quick Export** page will change accordingly. This is useful if you want to, for example, upload your videos in 4K rather than 1080p.

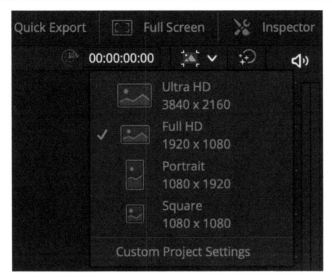

Figure 1.16: Timeline Resolution drop-down menu

We now know how to publish videos directly to three commonly used social media accounts directly from Resolve. However, we can also export our videos and upload them separately to other social media accounts. Let us look at how to do that next.

Preparing a video to upload to social media outside Resolve

Of course, there are more social media accounts than just YouTube, Vimeo, and Twitter. The list keeps on growing, along with their different formatting and aspect ratio requirements. Thankfully, in Resolve, you can quickly reframe and repurpose existing videos to use for different social media platforms, both current and future ones.

For example, let's take the video you have just edited and prepared for uploading to Instagram. According to Facebook, which owns Instagram, their video requirements are as follows:

"*Video Resolution and Size*:

- *You can upload a video with an aspect ratio between 1.91:1 and 9:16.*

- *Videos should have a minimum frame rate of 30 FPS (frames per second) and minimum resolution of 720 pixels.*

- *The maximum file size for videos that are 10 minutes or less is 650MB. The maximum file size for videos up to 60 minutes is 3.6GB.*"

Of course, these requirements can always change, so it is best to check for yourself on a regular basis to be sure.

Source: www.facebook.com/help/instagram/1038071743007909/?helpref=uf_share

Duplicating your Timeline

First, let's create a copy of our **Timeline** so that we can make changes to it without losing our original edit:

1. In the **Media Pool**, right-click on your **Timeline** and select **Duplicate Timeline**.

2. Rename the duplicated **Timeline** to something suitable, such as IGTV Edit, by performing the following:

    *   Click on the title below the thumbnail by clicking on it twice and entering a new name

    *   Alternatively, if in the **List** view, type directly into the **Clip Name** field

3. Double-click on the new **Timeline** thumbnail or icon (if in the **List** view) to open the new **Timeline**.

Now that we have a copy of our **Timeline**, let's reframe our video for the new **aspect ratio**.

Checking your content for a different aspect ratio

First, let's check our content for our new video's **aspect ratio**:

1. Click on the **Safe Area** button in the upper-left corner of the **Viewer** (*Figure 1.17*):

Figure 1.17: The Safe Area button

2. You will see a **Social Media** heading under which you have the following options:

    - 1:1

    - 4:5

    - 9:16

    - 1.91:1

    - 16:9

3. Select the one you prefer. In this case, we are going to select **9:16** as that is a popular format for Instagram TV (IGTV) and can also be used for Facebook and TikTok.

4. You will now see a white box on the **Viewer** that shows the safe area of our content for it to be seen in the 9:16 vertical format.

5. Scroll through the video on your **Timeline**, and you should notice that some of your content will suit the new frame; however, some will not.

You can either use **Ripple Overwrite** to replace the shots that are not framed well for the new **aspect ratio** or you can reframe the existing footage to suit.

Let's change the framing of the existing clips.

Reframing our video using the Transform controls

Let's reframe our video to suit the new aspect ratio:

1. Click on the **Tools** button (*Figure 1.18*) in the bottom-left corner of the **Viewer**. This reveals additional tools that we can use to enhance each clip:

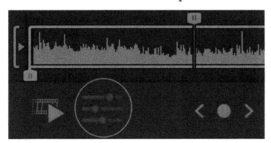

Figure 1.18: The Tools button

2. Select the **Transform** tool (the first button; see *Figure 1.19*). It will be underlined in red to show it has been selected, and the transform controls will be shown underneath.

3. Click and drag the **X** position control (the third transform control from the left) to change the **X** position value of the clip to move it to the left or right until the subject of the clip is inside the white **Safe Area** box:

Figure 1.19: The Transform button and controls

> **Top tip**
>
> To see before and after any effect, you can toggle the effect on and off by clicking on the red switch to the left of the controls. Anywhere you see this red toggle switch, you can use it to preview what an effect will look like on the video or even disable an effect without deleting it.

Feel free to go through each clip and use the **Transform** tool to reframe your video to suit your new aspect ratio.

Now, let's crop your video to suit the new aspect ratio by changing the **Timeline Resolution**.

Changing the timeline resolution

We now need to change your **Timeline Resolution**, which will also change the aspect ratio:

1. Click on the **Timeline Resolution** button (*Figure 1.16*) in the top-right corner of the **Viewer** to see a drop-down list of common **Timeline** presets:

    -   **Ultra HD 3840 x 2160**

    -   **Full HD 1920 x 1080**

    -   **Portrait 1080 x 1920**

    -   **Square 1080 x 1080**

2. Select **Portrait 1080 x 1920** for our 9:16 vertical format.

    You will notice that the video has now been resized down to fit the 9:16 vertical aspect ratio, but Resolve has added black bars above and below the original footage. We don't want this, but it is an easy fix:

    -   Select **Custom Timeline Settings** from the bottom of the **Timeline Resolution** drop-down list

    -   Under the **Format** tab, next to the **Mismatched Resolution** options, change the option selected from **Scale entire image to fit** to **Scale full frame with crop**

Resolve has now rescaled and cropped the video on the **Timeline** to suit the new aspect ratio. It is now time to export our newly reframed video ready to be uploaded separately to IGTV.

Exporting video using Quick Export

Quick Export can be used not only for directly uploading your videos to social media, but also to quickly export videos to your computer.

Quick Export has three main Video Codecs presets, H.264, H.265, and ProRes.

- **ProRes** is a Video Codec mainly used by professional video editors to edit with, as it has a bigger size and is less compressed to retain quality, but due to its large size, it is not so great for social media.

- **H.264** is a universally accepted Video Codec that efficiently compresses video into smaller file sizes while retaining quality.

- **H.265** is a newer version of H.264, so is not as universally accepted. But it is more efficient at compressing video for the same quality as H.264, so the file sizes can be smaller.

Let us choose H.264 for our social media uploads:

1. Click on the **Quick Export** button.
2. Select **H.264**.
3. Click on the **Export** button.
4. Navigate to a folder on your computer where you want to save your video.

   By default, Resolve will name your video after the **Timeline** name. However, you can rename your video before saving it to your computer.

5. Click on the **Save** button to save the file to your computer.

You have now repurposed an existing edited video for another social media platform by recomposing shots, cropping the video, changing its aspect ratio, and exporting it to your computer ready for separate upload at another time.

> Top tip
> You can add more presets to the **Quick Export** menu by creating them on the **Media** page. We will cover this in a later chapter.

Summary

Here is what you have achieved in *Chapter 1*:

- Learned about pages in DaVinci Resolve and how to hide them
- Saved your UI as an importable preset

- Imported footage into bins, and added, deleted, and exported bins
- Edited a rough cut using Track 1-specific editing tools such as **Smart Insert**, **Append**, and **Place on Top**
- Added cutaways using Track 2-specific editing tools such as **Close Up** and **Place on Top**
- Set up your social media accounts ready for publishing with Resolve
- Published your video to social media all within Resolve and not even leaving the **Cut** page
- Repurposed existing edits for different social media formats, reframing the footage using the **Transform** controls and adjusting the **Timeline Resolution** to change the **aspect ratio**
- Used **Quick Export** to create a copy suitable to upload to other social media accounts outside Resolve

You now know how to set up DaVinci Resolve to simplify the UI and use the editing tools on the **Cut** page to efficiently put together a rough cut and publish it directly to social media to save you time for your current and future video edits.

In *Chapter 2*, we will make the video you edited in *Chapter 1* look even more amazing by adding titles, transitions, and motion graphics, to add more visual interest on top of your existing footage. Additionally, we will look at how your can share your work in progress with others in your team so that they can help you with your edit.

Questions

1. True or false? Adding new clips to Track 1 on the **Cut** page deletes existing footage on the **Timeline**, whereas adding new footage to Track 2 ripples the footage on the **Timeline**.
2. What social media accounts can you publish videos to directly from Resolve?
3. Which editing tools work specifically with Track 1 of the **Timeline** on the **Cut** page?
4. Which editing tools work specifically with Track 2 of the **Timeline** on the **Cut** page?
5. True or false? *Timecode* is the term used to program time-based effects in DaVinci Resolve.

Further reading

- Oscar-nominated films in 2021 that used DaVinci Resolve or Fusion: `https://www.blackmagicdesign.com/uk/media/release/20210422-01`
- Top Sundance films and TV series using DaVinci Resolve: `https://www.blackmagicdesign.com/media/release/20190129-02`

- *Indian Tamil-Language Film 'Soorarai Pottru' Graded Using DaVinci Resolve*: `http://www.content-technology.com/postproduction/indian-tamil-language-film-soorarai-pottru-graded-using-davinci-resolve-studio/`

- Resolve used to Grade Sundance films: `https://magazine.artstation.com/2019/02/behind-the-scenes-sundances-top-films/`

- *Chinese Blockbuster The Eight Hundred Graded by Zhang Gen Using DaVinci Resolve*: `https://www.cinematography.world/chinese-blockbuster-the-eight-hundred-graded-by-zhang-gen-using-davinci-resolve-studio/`

- *Summer 2019's Biggest Film Releases Used Blackmagic Design*: `https://www.nexttv.com/post-type-the-wire/bmd-2019-summer-films`

- *Fusion Studio Used by Flash Film Works To Composite Fox's 'Spy'*: `https://indieshooter.com/fox-spy-blackmagic-fusion-studio/`

- *The Long List Of Summer Blockbusters Using Blackmagic Design Products*: `https://indieshooter.com/new-site/blackmagic-design-summer-movie-products/`

- *Summer movie lineup benefits from Blackmagic Design*: `https://www.panoramaaudiovisual.com/pt/2021/08/30/summer-movie-lineup-blackmagic-design/`

- Behind the scenes of creating the soundtrack for *Fight Club*: `http://filmsound.org/studiosound/pp_fightclub.html`

2

Adding Titles and Motion Graphics

In this chapter, you will open the project created in *Chapter 1* and learn how to share your project with other computers so that other people can also work on it. You will also enhance your video by adding a basic title and motion graphics using **Fusion FX** and transitions on the **Cut** page. You will also save your favorite effects and transitions to access them quickly for other projects.

In this chapter, we're going to cover the following main topics:

- Opening an existing project and exporting and importing it to use on other computers or share with others
- Adding a title and customizing it
- Adding transitions and customizing them
- Adding Fusion effects and customizing them
- Saving your favorite titles, transitions, and effects to access quickly later

Technical requirements

For the exercises in this chapter, we will be continuing to work on the project we created in *Chapter 1*. If you have not completed *Chapter 1*, then I suggest you quickly follow the exercises there so that you can get the most out of this chapter.

Opening, saving, and sharing projects

Before we start adding titles and effects to our video, let's look at how we can share our project with others so that they can work on it, or even just so that we can open it on another computer.

Whenever we need to save, archive, import, or export a project, we do this in the **Project Manager** (*Figure 2.1*).

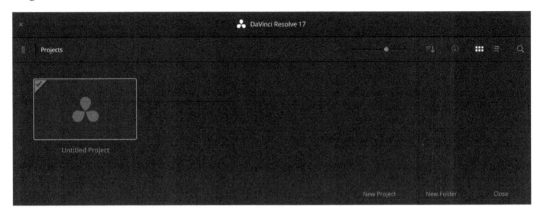

Figure 2.1: Project Manager

In the Project Manager, there are various ways to share our project outside of our computer, each with its own strengths and weaknesses. Let us look at each in turn.

Exporting a project

In *Chapter 1*, we looked at how to create and save a new project, but we did not look at how we can access the project to share it with others. You may have noticed that when we saved our project, it didn't ask us where on our computer to save it to. That is because Resolve saves our project into a database containing all our projects, rather than saving it as a discrete file.

Thankfully, Resolve does let us export individual projects to share with others rather than sharing our entire database. This is all done in the Project Manager. The Project Manager is where we manage all the creating, naming, importing, exporting, and archiving of our projects. Let us go there now:

1. If you are not already in the Project Manager, click on the **Project Manager** button (looks like a house, *Figure 2.2*) at the bottom-right of the screen below the **Timeline**.

Figure 2.2: Project Manager button

2. Select a project you want to export.
3. Right-click on the project and select **Export Project…**.
4. Select where on your computer you want to save your project and click the **Save** button.

You have now saved your project, which you can share with others to work on. You can recognize a DaVinci Resolve Project file on your computer as it uses the .DRP file extension. A DRP file only contains the project, which includes the bins and **Timeline**s, but it will not include the media – you will have to share this separately. You will then need to relink the project to the media stored on another computer. Project files are great if someone else already has access to the media on their computer, as the project files without the media are smaller and easier to share through email.

Exporting a project archive

If you do want to share your media with the project file, creating a project archive is the best way to do it:

1. If you are not already in the Project Manager, click on the **Project Manager** button (looks like a house) at the bottom-right of the screen.

2. Select a project you want to archive.

3. Right-click on the project and select **Export Project Archive...**.

4. Select where on your computer you want to save your project archive and click the **Save** button.

You have now saved your project archive, which you can also share with others to work on. You can recognize a DaVinci Resolve Archive file on your computer as it uses the .DRA file extension. A .DRA file contains the project file (.DRP), which includes the bins and **Timeline**s, as well as all the media you imported into your project. So, project archive files are great if someone else does not already have access to the media; however, they can be rather large in terms of file size, needing a separate hard drive to share them.

Importing a project

Of course, there is no point in knowing how to export a project if we don't know how to import one back in. Let us look at that now:

1. If you are not already in the Project Manager, click on the **Project Manager** button (looks like a house) at the bottom-right of the screen.

2. Right-click on an empty space in the Project Manager and select **Import Project...**.

3. Select where on your computer you have a saved project file, select it, and click the **Open** button.

4. Rename your new project in the **Name** dialog box.

A copy of your imported project will now appear in the Project Manager.

Importing a project archive

Similarly, there is also no point in knowing how to export a project archive if we don't know how to import one back in. Let us look at that now:

1. If you are not already in the Project Manager, click on the **Project Manager** button (looks like a house) at the bottom-right of the screen.

2. Right-click on an empty space in the Project Manager and select **Import Project Archive ....**

3. Select where on your computer you have a saved project archive file, select it, and click the **Open** button.

4. Rename your new project in the **Name** dialog box.

A copy of your imported project will now appear in the Project Manager complete with the media imported and linked up ready to continue working on.

The main difference between a project file and the archive file is that the project file does not contain media, whereas the archive does. So, although the archive file already has all the media linked, how do we do this for a project file if it does not already know where the media is located on your computer?

Well, the solution is easy in Resolve, and even easier in version 17 onward.

Relinking media (version 16 and earlier)

If your media needs to be relinked, it will show up as red on your **Timeline** and in your **Media Pool**, and a **Media Offline** warning message will show in your **Viewer**.

Figure 2.3: Version 16 Media Offline warning

Here is how to relink media in version 16 and earlier of DaVinci Resolve:

1. To relink the media, do one of the following:

    •   Select all the media that is red in your **Media Pool** and right-click it, and select **Relink Selected Clips…**

    •   Right-click the bin that the media is in and select **Relink Clips for Selected Bin**.

2. A popup will appear in which you can navigate to where the media files are on your computer. Locate the folder the files are in, then click **OK**.

3. If Resolve can't find all the files, it will ask you whether you want to do a deeper search of your computer. Click **Disk Search**, and Resolve will look on your entire computer for the media files. This option will take longer.

4. The media will now all be relinked, and the red **Media Offline** warning sign on the clips and in the **Viewer** will now be gone.

Relinking media (version 17 onward)

If your media needs to be relinked, it will show up as red on your **Timeline** and in your **Media Pool,** just like in Version 16 of DaVinci Resolve:

1. However, new to version 17 of Resolve is a **Relink Media** button (*Figure 2.4*) just below **Media Pool**; it will be red if the media is unlinked. Click this button.

Figure 2.4: The Relink Media button – looks like a chain link

2. A **Relink Media** popup will appear giving you the option of locating the media files on your computer or getting Resolve to do a deep search on your computer to find them:

    •   Click **Locate** if you know where on your computer your media is located, navigate to the folder location, select the folder, and click the **Open** button

    •   Click **Disk Search**, and select the disk you want Resolve to search if you cannot remember where the media is located. This option will take longer

3. The media will now all be relinked, and the red **Media Offline** warning sign on the clips and in the Viewer will now be gone.

Now, we know how to share our projects with others as either project or archive files and relink the media if necessary.

Let us add some transitions to our existing edited video (or someone else's edited video they have shared with you) and customize it.

Transitions – moving from shot to shot

Our existing edited video used straight cuts as transitions, and 90% of the time, this suits our purposes. However, there may be an occasion when you need to add another type of transition to help enhance your story, such as **Cross Dissolve** to show the passing of time.

On the **Cut** page, open up the project you were working on in *Chapter 1*, or if you want an extra challenge, import someone else's project so you can add transitions to theirs.

Key concept – transitions

A **transition** is basically how we move from one clip to another. Apart from the basic cut, all transitions consist of two clips that overlap each other slightly.

As we move from one clip to the other, the first one gradually disappears revealing the second clip underneath it. Transitions only vary in how they reveal the second clip underneath; for example, **Dissolve** slowly dissolves the first clip to reveal the second clip underneath it, whereas **Star** expands a star-shaped hole in the first clip until the second clip is fully revealed.

This is an important concept to understand as, if there is no footage to overlap (i.e., the edit points have no extra footage to draw upon), then the transition will not work. Resolve will show the edge of a clip as red if there is no room for a transition over that clip.

Adding a **Cross Dissolve** transition is extremely easy in the **Cut** page; there is even a dedicated button for it above the **Timeline**.

Transitions button (Cross Dissolve)

Let us add a **Cross Dissolve** transition to our **Timeline** now:

1. Place your playhead near an existing cut where you want the transition to appear (for example, a transition between scenes). You will see the **Smart Indicators** bouncing up and down showing you where the transition will go.
2. Select the **Dissolve** button (see *Figure 2.5*) above the **Timeline**.

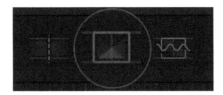

Figure 2.5: The Dissolve button

A Cross Dissolve has now appeared on your **Timeline**.

The **Dissolve** button is a quick way to add the default **Cross Dissolve** transition to your **Timeline**. However, if you want to have a more fancy transition, such as a Star Wars-style **Wipe** or an 80s TV ad **Starburst**, then you can find a variety of extras in the **Transitions** browser above the **Media Pool.**

Transitions browser

Let us add some additional transitions to our **Timeline** now:

1. Select the **Transitions** browser at the top-left of the screen near the **Media Pool.** You will see several different transitions under the **Video** tab, grouped under the following headings:

    - **Dissolve**: The various **Dissolve** options listed here differ based on how they dissolve to the image/clip underneath. For example, **Dip To Color Dissolve** dissolves to a solid color first before dissolving again to the next clip.

    - **Iris**: You can choose different shapes that open up (like a camera iris) to reveal the second clip underneath.

    - **Motion**: Here, you can choose how the first clip moves out of the way to reveal the second clip (or the other way around). For example, using **Slide**, the second clip slides over the first clip to totally replace the first clip.

    - **Shape**: Just like **Iris** but with additional shapes that open up to reveal the second clip.

    - **Wipe**: This is where the first clip is wiped away by the second clip. For example, with **Edge Wipe**, the second clip, starting from one edge, progressively replaces (wipes over) the first clip.

    - **User**: This is where you can find your own transitions that you saved after customizing them. We cover how to do this later in the chapter.

    - **Fusion Transitions**: These are advanced transitions using more complex animations created in Fusion. For example, **Zoom In** zooms in on the first clip before dissolving to reveal the second clip. The main advantage of **Fusion Transitions** is that they can be further edited in Fusion by Fusion artists.

- **Resolve FX Color**: This uses effects from the **Color** page. At the moment, there is only one transition, called **DCTL Transition**, which uses **DaVinci ColorSpace Transform Language** (**DCTL**) to transition between the clips. This is obviously only for advanced users.

- **Resolve FX Stylize**: This uses graphic effects to enhance the transition, however, they cannot be edited on the **Fusion** page as they are not Fusion transitions (they are Open FX instead). At the moment, there is only one transition, called **Burn Away**, where the first clip burns away using an animated fire effect revealing the second clip.

2. Choose a transition that suits the story you are telling. Less is more, unless you are trying to emulate the style of a particular television time period, such as 70s wipes or 80s starbursts.

3. Drag the transition directly onto your edit point (the point where two clips meet) on either the upper or lower **Timeline**.

You will now have the transition you just selected on your **Timeline**. Let us add it as a favorite transition so that we can quickly find it again.

Saving your favorite transitions

Scrolling down a long list of transitions to be able to find your favorite one is time-consuming, so let's see how we can mark transitions as favorites to be able to find them quicker in the future.

Let us add a regularly used transition as a favorite:

1. Select the **Transitions** tab at the top-left of the screen near the **Media Pool**.

2. Click on the **Star** icon to the right of the transition you want to save to **Favorites**. The **Star** icon will highlight white to show it is now saved as a favorite.

You will now notice a copy of this transition is added to a **Favorites** section in your **Transitions** tab. You should now be able to find this transition quickly should you ever need to use it again.

Wouldn't it be great to adjust a transition and then save the changed version to the **Transitions** tab to use later?

Well, we can do that in Resolve too.

Customizing transitions

Now that we now know the variety of ways we can apply a transition to our **Timeline**, let us now look at how we can customize the duration and effect of our transition:

1. On the **Timeline**, select the transition that you want to change. It will highlight orange to show that it has been selected.

2. Select the **Inspector** tab at the top-right above the **Viewer.**

3. You will see options to change the **Video** or **Audio** transition settings.

4. Under the **Video** settings, you will see the following universal options:

    •   **Transition Type**: Here, you can change the transition by selecting another transition in the drop-down list. These are the same transitions that are available in the **Transitions** tab.

    •   **Duration**: **Seconds** and **Frames:** This is how long the transition lasts in total. You can change **Seconds** or **Frames** depending on what you prefer. Changing your frames will be more precise due to it being based on your **Timeline's** frame rate, as the seconds can only be changed in 0.1 increments (i.e., 1/24th of a second based on 24 FPS is a more precise measurement than 1/10th of a second).

    •   **Alignment**: This is how the transition aligns between the two clips – covers the clip on the left, covers both clips, and covers the clip on the right.

    •   **Ease**: This changes the timing for a gradual start (**Ease In**) or end (**Ease Out**) of the transition, or both the start and end (**Ease In & Out**).

    •   **Transition Curve**: This is for adjusting the **Ease Custom** setting by keyframing the **Ease** animation. This is a more advanced setting and is only for those who want to have total control of the timing of their transition. As this book is for beginners, it is beyond the scope of this book.

    •   Other settings will also be available that are specific to the transition that is being applied.

5. Change some of the preceding settings to suit the style of your video, such as applying **Ease In & Out** and setting **Duration** to **3 seconds** to give a smoother more film-style look.

6. Click on the **Inspector** tab to close the **Inspector**.

Now that you have changed the settings of a transition, let us save this new altered transition as a user-defined transition in the **Edit** page so we can use it again later for other videos.

Key concept – the Inspector

The **Inspector** window is available on every page in Resolve apart from the **Color** and **Deliver** pages. The Inspector gives you access to more advanced settings for any element you have selected on the **Timeline**, such as titles, transitions, or clips.

You can also toggle on or off different settings available in the **Inspector** window by clicking on the red toggle switch next to each setting. This is great if you want to see the before and after results of a setting.

Reset arrows at the far right of each setting will change the setting back to its default settings in case you made too many changes.

Saving customized transitions (Edit page feature)

Let us save this adapted transition so that we can use it again in the future:

1. Select the **Edit** page.

2. Right-click on the transition on the **Timeline** you want to save.

3. Select **Create Transition Preset**.

4. Name your new transition preset in the **Enter Preset Name** dialog box and click the **OK** button to save it.

You have now changed a transition on the **Timeline** to suit your video footage and stored it for later use.

You can now find this transition on the **Cut** page in the **Transitions** tab: **Transitions** > **User**. It is also on the **Edit** page under the **Effects Library** > **Toolbox** > **Video Transitions** > **User**.

Now that you have added all these transitions, you may be thinking: "*How do I remove the transitions if I just want to go back to the basic cut?*"

Removing transitions

There are two approaches to removing transitions. After selecting the transition on your **Timeline**, you can do either of the following:

- Press the *Delete* key on your keyboard
- Click the **Cut** button (*Figure 2.6*) above the **Timeline** next to the **Dissolve** button

Figure 2.6: Cut button

Both of these approaches will have the same result: removing the transition and reverting back to a straight cut.

You can also remove several transitions at once by doing the following:

1. *Command + click* on several transition points on your **Timeline**, in any order you choose.

2. Then do either of the following:

   - Press the *Delete* key on your keyboard
   - Click the **Cut** button (*Figure 2.6*) above the **Timeline** next to the **Dissolve** button

You have now removed a number of transitions at once ready to apply some different ones.

Your video is now flowing a bit better between each shot to help tell the story.

Now, let us introduce the subject of our video by adding a title to our existing edited video (or someone else's edited video) and customize it.

Adding a title and changing it

On the **Cut** page, open up the project you were working on in *Chapter 1*, or if you want an extra challenge, import someone else's project so you can add titles to theirs.

There are many different types of titles:

- Some appear on a black (or a solid color of choice) background before any clips are seen
- Some appear over the top of clips where we can see the footage behind the title
- Some are called **Lower Thirds**, and appear on the bottom of the screen, introducing the name of the speaker or video location
- Some are called **Credits**, where all those who helped make the video are credited by having their name and contribution listed at the end of the video

Whichever title you choose (you are likely to use more than one), adding a title is extremely easy on the **Cut** page; there is even a special tab for it called **Titles**.

Let us add a title to our **Timeline** now:

1. Select the **Titles** browser at the top-left of the screen near the **Media Pool**.
2. A list of possible titles will appear:

    - **Titles** are basic titles that you can use. They are good for low-performing or older computers
    - **Fusion Titles** are title templates using Resolve's powerful animation and compositing tool, Fusion

3. Select the title you want and drag it down to your **Timeline**:

    - If you drag it onto Track 1, then the rest of the clips will ripple to make way for the title
    - If you drag it onto Track 2, then the title will be overlaid on top of the footage on Track 1

You should now have a blank title on your **Timeline**. Let us now look at the differences between basic titles and Fusion titles.

Basic titles

Standard Titles are less graphically intensive as they have fewer options, so are good for older, less powerful computers. They are also good if you want to do something basic very quickly or create temporary (temp) Titles for the client while someone else is creating them in Fusion. They will also play back smoother on the **Timeline**. Let us look at the options the standard Titles provide:

1. Put a regular Title on your **Timeline** as described previously.

2. Click on **Inspector** to reveal the Title's options.

3. In the textbox labeled **Rich Text**, type over **Sample Text** to add your own Title (e.g., `My First Film`). You will be able to adjust the look of your Title text by changing any of the following **Rich Text** settings (common to any word processing software):

   - **Font Family**: This drop-down list will reveal all the fonts available that have been installed on your computer.

   - **Font Face**: Depending upon the font selected, you will also be able to select whether you want it to be **Bold**, **Italics**, and so on.

   - **Color**: Here, you can use a color picker to change the color of your text.

   - **Size**: Font size slider.

   - **Tracking**: Size of space between each letter.

   - **Line Spacing**: Amount of space between each line of text.

   - **Font Style**: Has buttons to apply underline, overhead line, strikethrough, superscript, and subscript styling.

   - **Font Case**: Drop-down list to select whether your text is **Mixed** (default setting), **All Caps**, **All Lowercase**, **Small Caps**, or **Title Caps**.

   - **Alignment**: Buttons to align the text against an edge – for example, **Left**, **Center**, **Right**, or **Justified** (both left and right edges).

   - **Anchor**: Buttons to select how text is anchored to the current position of the text in the frame, which by default is the center of the frame. You can adjust the anchor both horizontally (top, centered, and bottom) and vertically (right, centered, and left).

   - **Position X, Y**: This adjusts the center position of the block of text. The **X** value moves the block of text left and right. The **Y** value moves the text up and down. The bottom left corner of the frame, in X, Y coordinates, is (0,0). Position controls are also found in the **Tools** button below the **Viewer** (more on this later).

- **Zoom X, Y**: This adjusts the **Zoom** position of the block of text. The **X** and **Y** values are linked together by default. The value of **1** is the original size of the text as set under **Font Size**; values higher than **1** are zoomed in, and values lower than **1** are zoomed out. You can click the **Link** button (small chain symbol between the **X** and **Y** values) to unlink the **Zoom** values. Unlinking the **Zoom** values enables you to stretch or squeeze the text vertically (**Y**) or horizontally (**X**). Zoom controls are also found in the **Tools** button below the **Viewer** (more on this later).

- **Rotation Angle**: This slider rotates the text 360 degrees around the center of the textbox. **Rotation Angle** controls are also found in the **Tools** button below the **Viewer** (more on this later).

As you can see, there are a number of ways to adjust the **Title** text in **Inspector**, but before we look at the options for **Stroke**, **Drop Shadow** and **Background**, let us look at other ways to adjust the position, zoom, and rotation of the **Title** text without using the **Inspector**.

Using the Tools button and on-screen text controls

Underneath **Viewer**, there is a **Tools** button (*Figure 1.24*), which, like the **Inspector**, reveals extra tools to adjust any element on the **Timeline**. We used the **Tools** button in *Chapter 1*, to reveal the **Transform** controls to resize our video footage.

In this case, we will look at the tools available to move the **Title** text:

1. Make sure you have the Title you want to adjust selected. It will be highlighted in orange to show it is selected.

2. Click on the **Tools** button under **Viewer** at the bottom-left next to the **Fast Review** button.

3. This will reveal different tools under **Viewer** that you can use to make basic changes to any **Timeline** element that you have selected. Let us look at the first set of tools, called the **Transform** controls (*Figure 2.7*); these can also be found in the **Inspector** under the **Settings** tab):

   - **Zoom Width and Height**: These first sliders (**Zoom Height** and **Width**) are the **Zoom** sliders and operate the same way as in the **Inspector**. Clicking the **Padlock** icon unlinks the Zoom X and Y values, enabling you to change the aspect ratio of the text.

   - **Position X**: This slider moves the text left and right along the *X* axis.

   - **Position Y**: This slider moves the text up and down along the *Y* axis.

   - **Rotation Angle**: This slider rotates the text around the center of the text.

   - **Pitch**: This slider changes the vertical angle of the **Title** text (just like the intro titles in the Star Wars films).

   - **Yaw**: This slider changes the horizontal angle of the **Title** text.

   - **Flip Horizontal**: This button flips the text horizontally.

   - **Flip Vertical**: This button flips the text vertically.

As usual, you can undo all of the changes you made by clicking on the *undo* arrow on the far right. However, you can also undo each individual parameter by clicking on its icon just before the slider.

If you have made any changes, you will see a little orange dot at the top-right of the **Transform** tool icon.

4. Click on the **Tools** button under **Viewer** again to hide the **Tools** controls.

You will have noticed that when you adjusted any of the **Transform** controls, revealed by the **Tools** button, a white bounding box (*Figure 2.7*) appeared around the edge of the **Viewer.** You can also use this to adjust the **Position**, **Zoom**, and **Rotation** settings of the frame that contains the whole title.

Figure 2.7: Frame bounding box and Transform controls

Independently of the **Transform** controls, you can also change **Position**, **Zoom**, and **Rotation** of the **Title** text directly in the **Viewer**. Without needing to use the **Tools** button, click directly on the **Viewer** and use the same on-screen controls to move the text bounding box (*Figure 2.8*).

Figure 2.8: Title text bounding box on screen resizing controls

Here are the options available when using the on-screen controls of whichever method you choose:

- **Position**: Click within the box and drag to move the text position
- **Rotation Angle**: Click on the small square (text controls) or circle (transform controls) above the vertical line coming from the top of the textbox and drag it to the left or right to rotate the text
- **Zoom**: Click on a corner and drag to zoom in or out. This will keep the X and Y dimensions of the text linked
- **Zoom Height**: Click and drag on the small box or circle on the top or bottom edge of the bounding box to change the zoom height
- **Zoom Width**: Click and drag on the small box or circle on the left or right edge of the bounding box to change the zoom width

> **Note**
> The **Transform** controls in the **Tool** button (and **Inspector**) and on the on-screen **Viewer** controls (and **Rich Text** position controls) operate independently of each other.

This means resetting one will not reset changes in the other. This is because the **Transform** controls change the position and size of the whole frame that contains the title, whereas the **Text** controls in the **Inspector** and the on-screen **Viewer** controls affect only the box directly around the text itself.

Now that we have mastered the on-screen **Viewer** controls to resize or transform our title, we can now return to the **Inspector**, where we will look at the options to help our title text stand out from its background.

Stroke controls

In the **Inspector** tab under the **Video** then **Title** tabs, below the **Rich Text** settings, you will find the **Stroke** controls. So, the pathway is **Inspector**> **Video**> **Title**> **Rich Text**> **Stroke**.

If you cannot find the **Stroke** controls, you will first need to click on **Rich Text** to reveal the **Rich Text** controls then click on **Stroke** to reveal the **Stroke** controls. There are only two **Stroke** controls:

- **Color**: Clicking on the color swatch gives you options to change the color of the outside edge (stroke) of each character of your text
- **Size**: This slider changes the size of the stroke, making it thicker or thinner (measured in pixels)

You can use the **Stroke** controls to change the outline border of your text to make it stand out from the background, particularly if it is overlaid over a video image.

Another way to help your title be seen against a video background is to add **Drop Shadow**.

Drop Shadow controls

Beneath the **Stroke** controls, you will find the **Drop Shadow** controls. If you cannot see the controls, click on the **Drop Shadow** heading and it will reveal the controls:

- **Color**: Clicking on the color swatch gives you options to change the color of the shadow of each character of your text.
- **Offset**: This slider and value box changes the position of the shadow relative to the text. The default is **0** for both **X** and **Y**, which shows no shadow:
  - **Offset X**: Negative **X** values make the shadow appear from the left, and positive **X** values make the shadows appear to the right
  - **Offset Y**: Negative **Y** values make the shadow appear from below, and positive **Y** values make the shadows appear from above
- **Blur**: This slider changes the blur of the shadow to have a softer or more defined edge.
- **Opacity**: This slider changes how seethrough or transparent the shadow is on the image behind it. A value of **0** is completely transparent, and a value of **100** is completely solid.

If used well, **Drop Shadow** can subtly help the text stand out from the video background without being too obvious.

Of course, an even more noticeable way to help your Title stand out from your video is to add a block background to mask the video directly behind the Title text.

Background controls

Beneath the **Drop Shadow** controls, you will find the **Background** controls. If you cannot see the controls, click on the **Background** heading and it will reveal the controls:

- **Color**: Clicking on the color swatch gives you options to change the color of the background box behind your Title.

- **Outline Color**: Clicking on the color swatch gives you options to change the outline color of the background box behind your Title.

- **Outline Width**: This slider changes the size of the outline, making it thicker or thinner (measured in pixels).

- **Width**: This slider changes the width of the background box.

- **Height**: This slider changes the height of the background box.

- **Corner Radius**: This slider changes the roundness of the background box's corners.

- **Center X and Y**: This slider and value box changes the position of the background box relative to the center of the text. The default is **0** for both **X** and **Y**.

- **Offset X**: Negative **X** values move the background to the left, and positive **X** values move the background to the right.

- **Offset Y**: Negative **Y** values move the background to the bottom; positive **Y** values move the background to the top.

- **Opacity**: This slider changes how seethrough or transparent the background box is on the image behind it. A value of **0** is completely transparent, and a value of **100** is completely solid.

The basic **Text Title** supports **Rich Text Format** (RTF) editing, which means we can change the formatting of individual letters and words in the Title as we would in any Text Editor.

There is another option to create Titles called **Text+**; this enables more options for animation, which is great for motion graphics. However, it does not support RTF editing so you can only apply one style to the whole block of text. Let us look at Text+ Titles in more detail.

Text+ titles

Although the **Text+** title is under the **Basic** titles section, it is actually a **Fusion Title** and as such, can be edited in Fusion, unlike the regular **Title**.

Let us look at some of the extra options the Text+ title gives us:

1. Put a **Fusion Title** on your **Timeline** as described previously.
2. Click on **Inspector** to reveal the Title's options.

In addition to the settings available for the basic title, a Text+ Title has additional options laid out under tabs to the right of the usual **Text** options (*Figure 2.9*).

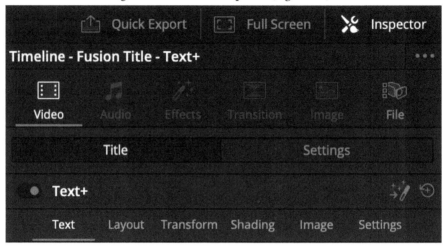

Figure 2.9: Text+ tabs

3. Clicking on a tab will reveal extra options to change your Title, and the tab will be underlined in orange to show that it is selected. The following is a sample of some of the tabs you will encounter:

-   **Text**: In addition to the same options you found with the basic Title, the Text tab has the following:

    -   **Direction**: The order in which each character on a line starts. This is great for Titles in Asian languages that read right to left.

    -   **Line Direction**: This is the order of each line of the title if it has more than one line. That is, the first line becomes the bottom line, and the title reads from the bottom of the page.

    -   **Write On**: This animates, using Key Frames, the start and end of the Title text appearing on the screen, so individual letters can appear one at a time.

    -   **Tab Spacing**: This controls the horizontal positions and alignment of eight separate tab stops, like in any word processor. If you copy and paste text from a word processor document that has tab positions in it, the text will align to these tab settings. This is useful to quickly create a two-column credits title.

-   **Layout**: This has settings that change the layout of the animated text (to a point, frame, circle, or path), rotate the text in a 3D space, or change the color of the background.

-   **Transform**: This has controls to change the spacing and pivot point of the title's characters, words, or lines. It also has controls to change the shear and size of the Title on the x and y axis, as well as the Title's rotation in a 3D space (i.e., X,Y,Z values).

- **Shading**: This has eight elements that change the text outline, border, shadow, and fill color of the title text. The first four elements (1 to 4) already have presets that you can change or use as they are. The last four (elements 5 to 8) are blank and ready for you to change and name them to suit. Like the **Transform** controls, there are **Shear**, **Size**, and **Rotation** controls, which control the position of the text outline or fill, shadow, or border, depending upon which element you have selected.

- **Image**: This is an advanced setting that changes the resolution, interlacing, gamma, and color space of each frame "image" of the Title.

- **Settings**: These are more advanced settings that allow you to apply masks and **Motion Blur** to your Title.

All of these **Text+** settings can be combined and animated using keyframes, which makes this a much more powerful tool compared to the usual Text Title generator. In reality, you will probably only change a select few settings to enhance your Title. Change one setting at a time to see the effect it has before changing the next one; as they say, "less is more."

> **Top tip – Slider Undo button**
>
> If an individual slider has its settings changed from the default value, a little white dot (reset button), appears under the slider as a visual cue to show that the setting has changed. See *Figure 2.10*, where the reset button has appeared under the text **Size** slider.
>
> If you get too carried away and want to just reset the last setting you have changed, you can click on this little white dot underneath the settings slider. This will reset that setting back to its default value. The default value is not always zero. With some effects, such as **Fusion FX**, the default value is a preset value for the effect as a starting point for when the effect is first applied.

Figure 2.10: Slider value reset button

Now that we have looked at some of the extra functionality that Fusion brings to the Text+ Title, let us look at dedicated Fusion Titles.

Fusion titles

Fusion Titles allow us to do a lot more, such as animate the Titles and create some simple motion graphics. They also have the advantage that they can be edited directly on the **Fusion** page to create even more advanced and customized effects.

You can recognize a Fusion Title (including Text+) on the **Timeline** by the small **Fusion** icon (*Figure 2.11*) to the bottom-left of the Title.

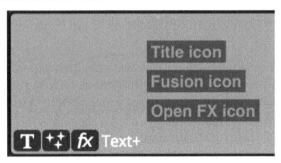

Figure 2.11: Fusion Title icon

Let us look at some of the extra options the Fusion Titles give us:

1. Put a Fusion Title on your **Timeline** as described previously.

2. Click on **Inspector** to reveal the Title's options.

    Most Fusion Titles will only allow you to change the basic aspects of the text, such as **Font**, **Size**, **Position**, and **Color**. The animation effects are all changed and controlled within Fusion. Think of the Fusion Titles as presets that are samples of the power you can uncap within Fusion.

    Some Fusion Titles will have limited additional options, usually to do with the specific effect that Fusion is applying to the Title.

    The following is a sample of some of these options you may encounter:

    *   **Large Text**: Change the format of the large text in an animated Title or lower thirds

    *   **Small Text**: Change the format of the small text in an animated Title or lower thirds

> **Key concept – Lower Thirds**
>
> Lower thirds are Titles that are placed in the lower third of the screen. They are usually used to display someone's name and job title or location of a scene. Of course, where you actually place them is up to you by using the **Transform** controls in the **Inspector**.

Now that you have chosen your Title, adjusted it, and helped it stand out from your background, just like transitions, you can save it as a favorite to use later.

Saving titles as favorites

Let us add a regularly used Title as a favorite. This is the same approach regardless of whether it is a basic Title or a Fusion Title:

1. Select the **Titles** tab at the top left of the screen near the **Media Pool**.

2. Click on the **Star** icon to the right of the Title you want to save to **Favorites**. The **Star** icon will highlight white to show it is now saved as a favorite.

You will now notice a copy of this Title is added to a **Favorites** section in your **Titles** tab. You should now be able to find this Title quickly should you ever need to use it again.

As you can see, there is some basic customization of Text for Titles. However, if you want to animate your Titles, it is better to choose a Fusion Title.

Top tip – saving titles as templates

Unlike transitions, there is no direct way on the **Cut** page to save your Titles for future use. To be able to save a Title directly in the **Title** tab, you need to open it in **Fusion** and save it as a Macro. This does require a more advanced understanding of Fusion, which is beyond the scope of this book and what most videographers have time to learn.

However, there is an easier way to save your Titles for future use. Go to the **Edit** page and drag the Title from **Timeline** into **Media Pool** and it will create a copy of your Title in **Media Pool**, which you can then drag onto any **Timeline**.

All Titles are recognizable on the **Timeline** by the **Title** icon in the lower-left of the clip (*Figure 2.11*).

Now that you have gained a better understanding of the differences between regular Titles and Fusion Titles and put them onto your **Timeline**, let us look at using some Fusion effects to enhance your video even more.

Adding visual effects

Resolve has two types of effects you can use to enhance your videos: **Fusion Effects** and **Resolve Effects**. Fusion Effects are Fusion templates created in Fusion and can be further edited in Fusion. Resolve Effects are plugins that use the **OpenFX** standard.

So you don't need to be a visual effects expert in order to apply some basic effects to your video.

Let us give an overview of the different types of pre-made effects available in Resolve. All of these effects can be applied by dragging them onto a clip on the **Timeline**. You can also add effects by clicking the **Add Effect** button that appears at the bottom of the **Effects** browser once it is selected (*Figure 2.12*).

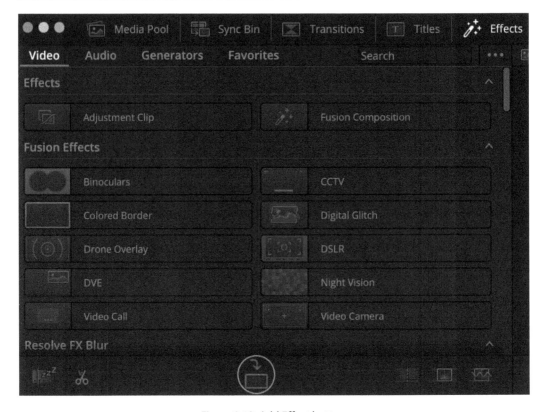

Figure 2.12: Add Effect button

Select the **Effects** browser tab at the top-left of the screen near **Media Pool**. You will see several different effects under the **Video** tab, grouped under the following headings (*Figure 2.12*):

- **Effects**: These are placeholder clips that are dragged onto the **Timeline** like any other video clip. You can also add effects by clicking the **Add Effect** button that appears at the bottom of the **Effects** browser. There are only two choices here:

  - **Adjustment Clip** allows you to make adjustments to it that will change the underlying tracks. This is good if you want to apply one effect across multiple clips on the **Timeline**, rather than apply the same effect separately to more than one clip. Or you could add an adjustment clip to apply an effect for only part of a clip, by making the adjustment clip shorter than the clip underneath. To do this, you simply apply (double-click or drag) any effect onto the adjustment clip rather than the clips underneath.

  - **Fusion Composition** is a clip that is edited in Fusion to create a more advanced composite than you can achieve in the **Timeline**. Great for adding your own Fusion Titles that you create directly in Fusion.

- **Fusion Effects**: As the name implies, these effects are pre-made in Fusion. You can adjust these effects in the **Inspector**, or for more advanced options, you can edit them directly in Fusion. To apply these effects, you can do either of the following:

    · Place your playhead over a clip on the **Timeline** and then double-click on a Fusion effect to apply it to the clip

    · Drag a Fusion effect directly onto a clip on the **Timeline**

    The **Fusion** icon (*Figure 2.11*) appears on the clip to show that a Fusion effect is applied to the clip.

The following effects are not Fusion effects but Resolve effects (based on the Open FX standard), which means that they cannot be edited in Fusion.

Resolve effects are applied to a clip the same way a Fusion effect is applied. When a Resolve (or Open FX) effect is applied to a clip, it shows an **Open FX** icon (*Figure 2.11*). You can also combine both Fusion and Resolve effects onto the same clip (or even a Title) and adjust them both in the **Inspector**:

- **Resolve FX Blur:** These effects apply various styles of **Blur** effects to the clips.

- **Resolve FX Color:** These are various preset **Color** effects borrowed from the **Color** page. More useful for professional color processing without needing to use **Resolve Color Management** on the Color page.

- **Resolve FX Generate**: These effects generate images that interact with the footage underneath:

    · **Color Generator** generates a single color that can be used as a color filter over the footage by adjusting the blend of the color chosen with the image below.

    · **Grid** creates a customizable grid overlay.

    · **Color Palette** (studio version only) creates a color palette overlay from the key dominant colors in the video footage. This is good for analyzing the colors of a scene and is hence great as a starting point for creating graphics that tonally match the footage.

- **Resolve FX Key**: These effects help remove ("key" out) backgrounds and create video layers on top of each other ("composites"). The Keyers only differ in the way that they select the background to be able to remove it to make that part of the video transparent to the video layer underneath:

    · **3D Keyer** removes color, such as from a green or blue screen.

    · **HSL Keyer** (short for Hue, Saturation, and Luminance) uses a mixture of **Color** (hue), **Color Strength** (saturation), and **Brightness** (luminance) to key out the background.

    · **Luminance Keyer** only uses the luminance of an image, not color, to remove the background. For example, this can remove a bright blue sky over a dark blue ocean, whereas a color-based key would not be able to do this successfully because the colors are too similar.

- **Resolve FX Light**: These effects mimic the way that light works with the atmosphere or the camera lens to give a particular look, such as **Lens Flair** (Studio version only) or **Light Rays** (also known to photographers as God Rays).

- **Resolve FX Refine**: The **Beauty** effect (Studio version only) allows you to soften the image, much like the Beauty filter photographers use, or in the early days of cinema, where they would smear Vaseline over the lens to soften the image when filming their "leading lady."

- **Resolve FX Revival**: These Studio version-only effects let you fix damaged or poor-quality footage.

- **Resolve FX Sharpen**: These Studio version-only effects give you detailed controls to sharpen specific parts of your video footage.

- **Resolve FX Stylize**: These effects add creative, artistic looks to your footage, such as **Mirrors** to create a Kaleidoscope effect in your footage.

- **Resolve FX Temporal**: These effects change the footage over time (temporal), such as using the **Stop Motion** effect to create a stop motion stuttering effect to live action footage.

- **Resolve FX Texture**: These effects add textures to your footage. For example, **Film Damage** recreates the look of an old-style damaged film.

- **Resolve FX Transform**: These effects can resize or move (transform) your video either by itself or as a layer over another video. This is good for picture-in-picture effects. We will cover these tools in a later chapter.

- **Resolve FX Warp**: These effects distort (warp) the image in different ways. For example, **Ripples** will create a water-like ripple through the footage, as if a stone has been thrown into a pond.

With all the above effects, you can add more than one effect to the same clip. For example, you could combine **Binoculars** and **Night Vision** Fusion effects to give the effect of night-vision goggles. You can also use Resolve and Fusion effects together.

You now have an overall understanding of the range of different types of effects available in Resolve. We will be showing you how to use specific effects from this range in more detail in later chapters. However, the best way to understand the wide range of preset effects available is to have fun playing around with them, adding one effect at a time and seeing what they each do.

For more advanced users, you can create your own effects in Fusion, or if you don't have time, you can purchase and import Open FX plugins into Resolve to add more effects to the effects already built into Resolve.

Hopefully, you now have a better understanding of the different options available to you to add titles, transitions, and effects to your video in Resolve.

Summary

Here is what you have achieved in this chapter:

- Learned how to open an existing project and export and import it to use on other computers or share with others
- Added a title and customized it
- Added transitions and customized them
- Added visual effects and adjusted them
- Gained an overview of all the effects available and their categories
- Saved your favorite titles, transitions, and effects to access quickly later

In *Chapter 3*, we will make the video you edited in *Chapters 1* and *2* sound better. We will do this by adding separately recorded sound, synching it with the video, and fixing and enhancing the sound with built-in and imported audio effects. We will also look at some basic sound theory, such as microphone choice and placement, to get the best quality sound in the first place.

Questions

1. True or false? Exporting a DRP file allows you to work on your Resolve project on another computer without the need to import any media files.

2. True or false? You can only have one font type with **Text** titles, whereas **Text+** titles allow you to use more than one font type in the title.

3. What is the default transition allocated to the **Transition** button? Choose the correct answer:

   A. **Star Burst**

   B. **Star Dissolve**

   C. **Star Wipe**

   D. **Happy Wipe**

   E. **Cross Dissolve**

4. True or false? Saving your favorite title, transition, or effect is as easy as clicking the little **Star** icon to the right of its name.

Further reading

More information about the **Open Effects (OFX)** standard that DaVinci Resolve uses for visual effects plug-ins, as well as a list of companies that support it and make their own plug-ins for it, can be found here: http://openeffects.org

3

Polishing the Camera Audio – Getting It in Sync

In this chapter, we will look at how recording separate audio can improve your video and how Resolve makes it easy to sync separate sounds up to your video. We will also look at how we can fix and enhance our audio quickly using **Fairlight** and AU Effects plugins.

In this chapter, we're going to cover the following main topics:

- Understand sound and whether to use an on-camera mic or separate mic to record audio and the best practices for recording it

- How to sync separately recorded audio to video footage

- Fix common sound issues using **Fairlight FX** on the **Cut** page

- Use **Voice Isolation** to fix audio in the **Inspector**

Technical requirements

To follow along with the fixing sound exercises in this chapter, download the following sound file and import it into a new project: `https://packt.link/B5bqz`

For these exercises, you will be fixing an audio-only file that you have placed on the **Timeline**. The techniques you will learn will apply equally to audio attached to video clips.

Understanding sound and different ways to record it

It is said that sound is the often-forgotten part of the filmmaking process. We often, at our own peril, believe that film is a visual medium and dedicate all our effort to getting the best picture possible, and then forget about sound until the last minute. However, video is both a visual and audio medium. Even the "silent" era of early film was accompanied by live music performed in the cinema to enhance the emotion of the story.

Without sound, a film or video is a lot harder to understand. Poor sound makes it unwatchable. You are more likely to switch off a TV program if the sound is poor or unintelligible than if the visuals are technically poor. Whether it be the low-resolution images of *The Blair Witch Project* or the dimly lit scenes in *Game of Thrones*, we will watch these programs quite happily if we can hear clearly what is going on, even if we can't see it.

At the very least, if sound is 50% of the video experience, then we should dedicate a bit more time to it.

If we start from the shared understanding that sound is important and that it should not be an afterthought, then let us look at what we need to know to capture the best-quality sound that we can before we even edit it in Resolve. To do that, we need to understand what sound is.

The properties of sound

Sound is essentially air pressure moving through the space around us as waves.

Just like waves on the sea rise and fall depending upon the amount of force behind them, sound works in the same way. To illustrate this, place your hand over the speaker of your phone or stereo when it is loud and full of bass, and you will feel the pressure of the air being pushed against your hand in time with the loudness of the music.

Let us look at the key terms used to describe the properties of sound:

- **Amplitude**: Like waves in the sea, we measure sound based on the height of the waves, a bit like sonic surfers. The taller the wave, the louder the sound (see *Figure 3.1 – A picture of sound waves*). The height of a sound wave is called amplitude. The word *amplifier* (or *amp* for short) that we use for our car or hi-fi stereos is based on the word *amplitude*. So basically, an amp increases the size (and loudness) of the sound wave.

- **Frequency**: Another way we measure sound is how often sound waves travel to us in the same time period. Using our sea example, if the waves of the sea only arrive at the beach every other minute, then it would be very calm. However, if several waves crash on the beach within a minute, then we would say it was time to get our surfboard out! How often sound waves travel to us is called *frequency*. A low frequency is fewer waves per minute and a high frequency means more waves per minute (*Figure 3.1 – A picture of sound waves*).

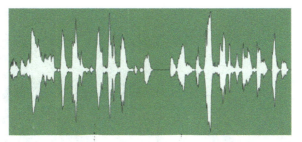

Figure 3.1 – A picture of sound waves

It is these sound waves that we see when we edit in Resolve. We use these waves to guide us on where to make the best cut. It is best to cut or trim a sound clip at the point when the wave is at its lowest point (i.e., there is minimal or no sound). Otherwise, we could be cutting someone off halfway through a word, and that would sound jarring to the listener.

Microphone types

Before we talk about which microphone to use, let us understand how a microphone records those sound waves. Regardless of the type of microphone, all microphones essentially work the same way. The air pressure of the sound wave vibrates a very thin material at the same frequency as the sound wave itself. This frequency is converted into an electrical signal that is then recorded as digital data on your video or sound tape/disc.

There are two common types of microphones you may come across, dynamic and condenser. The main way they differ is in how they convert sound waves into an electrical signal. Rather than go into the mechanics of how each record sound, let us look at what their strengths and weaknesses are:

- *Dynamic microphones* are used in loud concerts and stage work, as they are durable against knocks and bumps. Dynamic microphones do not need a power source to work, which makes them less sensitive to picking up different sounds but more durable.

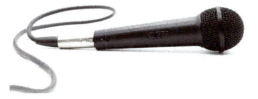

Figure 3.2 – A dynamic microphone (image by Juan José Berhó from Pixabay)

- *Condenser microphones* are usually used in film and video work, as they are also more sensitive to quieter sounds than a dynamic microphone, which does not perceive as much of the subtle sound of the spoken voice. Condenser microphones need a power source to work (such as a battery or a powered feed (*phantom power*) from a mixing desk), which makes them more sensitive but less durable. Condenser microphones can be used outdoors if handled carefully.

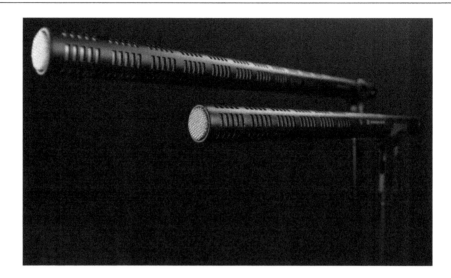

Figure 3.3 – A condenser microphone (source – Pexels)

In brief, a dynamic microphone is the least sensitive but most robust of the two, while the condenser microphone, in regard to its sensitivity and hardiness, is the opposite of the dynamic microphone.

Now that we know broadly about the different types of microphones, let's look at the pickup patterns of different mics in order to get the best possible sound.

Microphone pickup patterns

Knowing how sensitive a microphone is to sounds from different directions will help us know where to place it in regard to the sound source we are recording.

The direction from which a microphone records sound is called a *pickup pattern*. Pickup patterns are illustrated by a graph showing how sensitive the microphone is to the sound 360 degrees around it. The microphone is in the center of the graph and the sensitivity is measured in dB, where a value of 0 dB is the most sensitive to sound and a value of -25 dB value is the least sensitive. A black line on the graph shows the strength of the microphone's sensitivity to sound from different directions, with 0 degrees being the forward-facing direction of the microphone.

Let us look at the three most common pickup patterns and their variants.

Omnidirectional

Essentially, *omni* is a Latin word that means *all* (e.g., when people say "God is omnipotent," they mean God is all-powerful). So, when a microphone is labeled as *omnidirectional*, it means it picks up sounds from all directions 360 degrees around the microphone. These microphones are great when you want to record the atmosphere or background noise of a room and give a great sense of space.

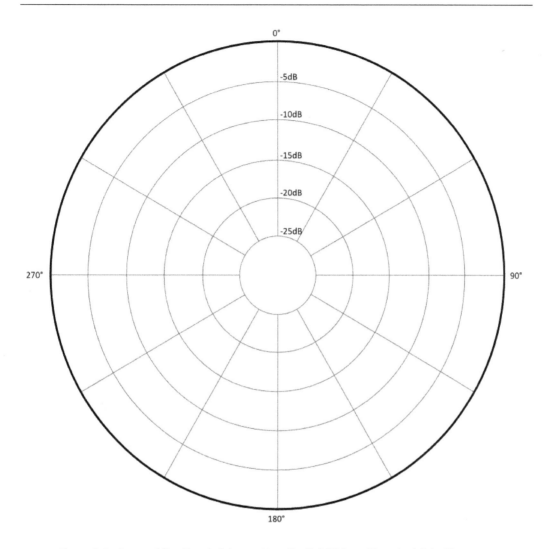

Figure 3.4 – An omnidirectional pickup pattern (by Galak76 – self-made, Adobe Illustrator,
CC BY-SA 3.0: https://commons.wikimedia.org/w/index.php?curid=2025973)

Unidirectional

Essentially, *uni* is a Latin word that means *one* or *single* (e.g., a unicycle has one wheel). Unidirectional microphones only pick up sound from one direction (not the UK boy band called One Direction). These microphones are great for recording just a specific sound, such as a voice, while eliminating sound from other directions. There are two main types of unidirectional mics based on the shape of their pickup pattern.

Cardioid

The pickup pattern of these mics is based on a heart shape; their name originates from the Greek word for *heart* (e.g., a cardiac arrest is a heart attack). This pickup pattern is the most sensitive directly in front of the microphone, reducing in sensitivity until directly behind the microphone where there is no sound pickup. Dynamic microphones use the cardioid pickup pattern to reduce the microphone handling noise from the performer or interviewer.

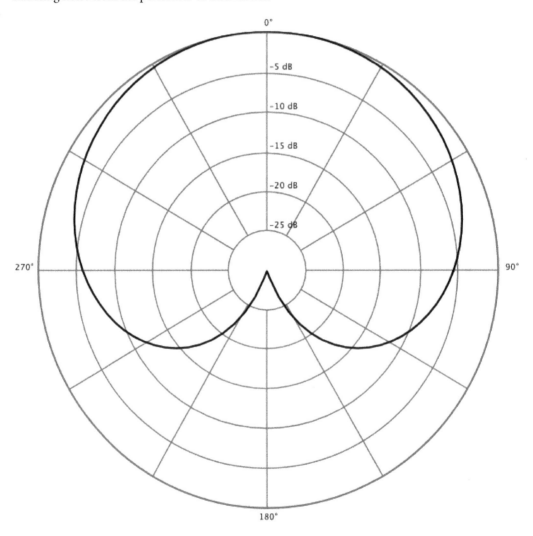

Figure 3.5 – The cardioid pickup pattern (by Nicoguaro – own work, CC BY 4.0: https://commons.wikimedia.org/w/index.php?curid=50230608)

Hyper-cardioid

Hyper comes from a Greek word that means *over* (e.g., *hyperactive* means *overactive*). Hence, this pickup pattern is a stretched or longer version of the cardioid pickup pattern. This pickup pattern is often used for **boom mics** (microphones held on a pole above the talent), as they are the most sensitive to unidirectional sound. However, they can pick up some camera and operator handling directly behind the microphone so need to be placed carefully.

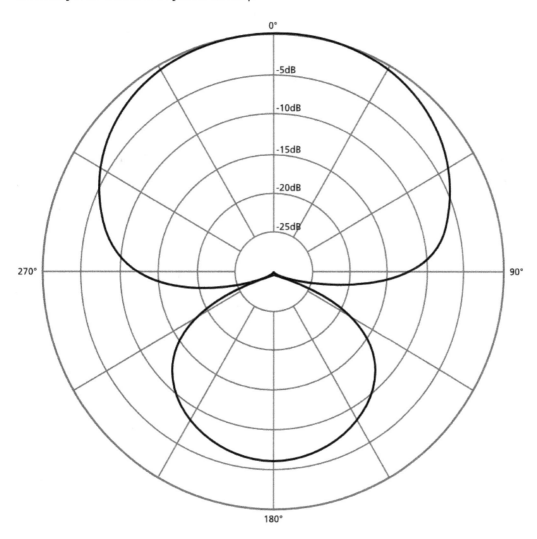

Figure 3.6 – The hyper-cardioid pickup pattern (by Galak76 – self-made, Adobe Illustrator, CC BY-SA 3.0: https://commons.wikimedia.org/w/index.php?curid=2025973)

Bidirectional

You guessed it! The word *bi* is Latin for *two* (e.g., a bicycle has two wheels). It is also called *figure of eight*, based on the shape of the pickup pattern. These microphones record sound from two different directions. These microphones are less used in filming but are good if you want to record a conversation between two people but only have one microphone to use. However, this is not a common approach, as using a bidirectional microphone to record two voices makes it harder to separate the voices when we are editing the sound. Also, each voice will have different vocal characteristics (e.g., quieter and louder), so it is better to have a separate microphone for each separate sound source so that we can adapt the microphone position and gain of each one to suit each sound.

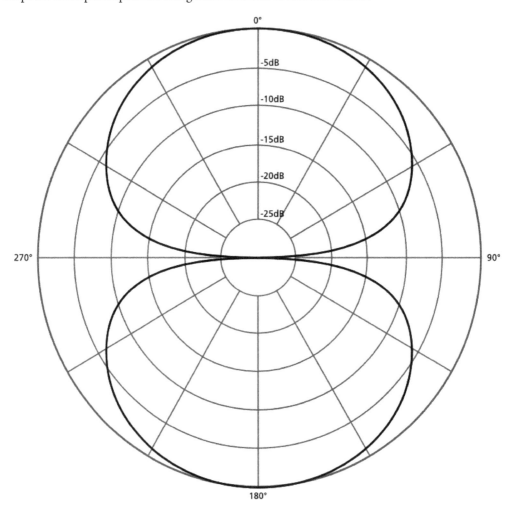

Figure 3.7 – Bidirectional pickup pattern (by Galak76 – self-made, Adobe Illustrator, CC BY-SA 3.0: https://commons.wikimedia.org/w/index.php?curid=2025973)

Most professional microphones have interchangeable heads to change a microphone's pickup pattern to suit what you are recording. Even some more affordable microphones have a switch to change their pickup pattern to suit.

Now that we know which microphone to choose for each particular job, let's look at microphone placement to get the best possible sound.

> ### Lapel (Lavalier) mics
>
> Lapel mics are tiny microphones that clip onto the lapel of the jacket worn by the person speaking. Lapel mics are not just named after a piece of clothing but also share the name of an item of jewelry called *Lavalier*, which is an ornament hanging from a chain and worn around the neck. This is because early Lav (Lavalier) mics were suspended from a chain/lanyard around the wearer's neck. Lavalier mics now have clips so that they can be attached to the lapel of a jacket or anywhere that can be easily concealed in a person's clothing, while remaining close (approximately 10 to 15 cm away) to their mouth.
>
> Most lapel mics are omnidirectional mics. It may seem strange that lapel mics are omnidirectional, as we only want to record the voice of the microphone wearer (subject) from one direction.
>
> However, as a person can change the direction of their head when they are talking, an omnidirectional microphone can pick up these changes successfully. Also, an omnidirectional microphone will use the body of the speaker to block unwanted sounds coming from behind.
>
> One thing to consider is clothing noise, as the subject moves. This can be eliminated by fixing a mic securely to the clothing so that it moves with it, not against it, or making sure that the clothing is tight enough that there is no excess movement against the mic.
>
> An advantage of lapel mics being omnidirectional for a sound editor is that they can also pick up the conversation of another person in a dialogue or conversation. This can then be used in an emergency if the other person's mic failed to record good-quality sound.

Microphone placement

The best mics to use are the ones closest to the sound source. Usually, this is the lapel mic, as being attached to the lapel of a speaker's jacket places the mic closest to their mouth. Lapel mics can be discrete and easily hidden within clothing.

You may have seen pictures of sound recordists using unidirectional boom mics to record voices in a film. These are great for when a scene is shot on closeup where the mic is placed just out of frame. However, they are no good in wider shots, as they are too far away and start to pick up more of the ambient noise around them. Most professional sound recordists for cinema use a mix of both boom and lapel mics to get the best of both worlds, and they can be used as backups for each other if one of them should fail. However, when filming a realistic scene or documentary, a boom mic and operator can get in the way of the natural performance of the speaker (or talent), as they are distracting and draw attention to the artificial nature of video recording, resulting in a self-conscious performance. Which you choose to use is, of course, up to you; these are just a few things to consider to help you make your choice.

Because of these considerations, there is a temptation for videographers to use the built-in microphone on a camera. The problem with this approach is that the microphone is usually too far away from the subject and closer to the sound of the handling and operation noise of the camera. Also, most cameras that have onboard mics use cheaper lower-quality mics to make the cost of the camera more affordable. The camera manufacturers tend to spend their research money on a better-quality image, as that is what you are probably buying a camera for in the first place. This all results in poor-quality sound coming from the on-camera microphone, which can be hard to improve upon in any editing software. Unless you are recording a video piece close to a stationary camera, it is much better to record sound separately on a lapel or boom mic closer to the subject. Even using a separate mic plugged into the camera with a very long lead to record directly into the video footage is better than using the on-camera mic.

One problem with long leads is that they can pollute the sound signal going back to the camera with noise. The longer the lead, the more likely it is to introduce this noise. This is usually why microphone cable runs are often as short as possible.

Also, microphone cables should never be run parallel to power cables for other equipment, such as lights or cameras; otherwise, they run the risk of picking up electrical hum.

The good news is that Resolve has special tools that can remove electrical hum or noise from poor sound, which we cover later in this chapter.

Now that we know the basics of different microphone types, the direction they record sound from, and their placement, let us look at how to get separately recorded sound into Resolve and synced up with our video footage.

Syncing audio

One reason that most fledgling filmmakers do not want to record sound separately is that they think that it is difficult to match the sound with the video they have recorded. The need to make the audio sync with the mouth movements of the video's subject seems daunting.

Thankfully, it is very easy to sync audio with video in Resolve. There are a few ways to do this, which we will look at now. Firstly, you will need to go to the **Media** page, as you can only sync audio on the **Media** page. If the **Media** Page is still hidden from when we hid it in *Chapter 1*, you can reveal it again by selecting it in **Workspace** > **Show Page** > **Media**.

Auto-syncing audio to video

Even though we will not use the sound recorded by the on-camera microphone, it is useful to help sync our separately recorded sound.

As discussed earlier in the chapter, sound is recorded as waves; the shapes of these waves are called *waveforms* (i.e., the shape or form of the wave). Resolve can take your separately recorded audio and match it up with your video's audio, based on the shape of the waveforms for both pieces of audio. All audio syncing is done on the **Media** page.

The **Media** page in Resolve is where you have more advanced functions to manage your Media, such as syncing audio to video. The **Media** page is not to be confused with the **Media Pool**, which is available on every page in Resolve but is limited to arranging media that has already been imported into Resolve.

Let us auto-sync audio based on a waveform on the **Media** page:

1. Go to the **Media** page.

2. Import both the separate audio and the video into the same bin in the **Media Pool** (as described in *Chapter 1*). The audio and video can be imported into the **Media Pool** on any page, not just the **Media** page.

3. To select the audio and video you want to sync, you can do one of two things:

    *   Select both the audio and video clips you want to sync. You can either click *CMD* (*Command*) or *Shift*-click on the clips to select them, or press the *CMD + A* keys on your keyboard to select all the clips in the bin if you want to sync them all at once.

    *   Select the bin that contains the video or audio you want to sync.

4. Right-click on any one of the selected clips or bins to reveal a pop-up menu (*Figure 3.8 – The pop-up menu*). Select **Auto Sync Audio**, as shown in the following screenshot:

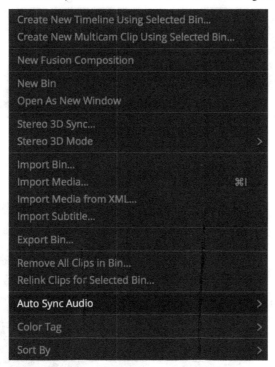

Figure 3.8 – The pop-up menu

This will bring up the following options:

- **Based on Timecode**: This option synchronizes the video and audio based on a matching timecode and replaces the video clip's existing audio with the newly synced audio files.

- **Based on Timecode and Append Tracks**: This option synchronizes the video and audio based on the matching timecode, and adds the newly synced audio files as additional audio tracks below the video clip's existing audio track. This means you can edit the original camera audio and the newly added synced audio separately and mix between them.

- **Based on Waveform**: This option synchronizes the video and audio based on matching waveforms and replaces the video clip's existing audio with the newly synced audio files.

- **Based on Waveform and Append Tracks**: This option synchronizes the video and audio based on matching audio waveforms and adds the newly synced audio files as additional audio tracks below the video clip's existing audio track. This means you can edit the original camera audio and the newly added synced audio separately and mix between them.

5. Your clips will now be synced.

    Whichever option you choose, a progress bar dialog box (*Figure 3.9 – The Auto Sync Audio progress bar*) will pop up to show you how long the syncing is taking.

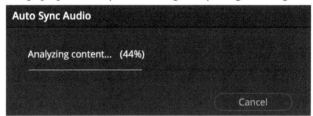

Figure 3.9 – The Auto Sync Audio progress bar

6. When the syncing is finished, the dialog box will list any video or audio clips that did not sync (*Figure 3.10 – The Auto Sync Failed list*).

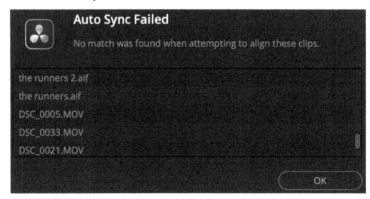

Figure 3.10 – The Auto Sync Failed list

Do not worry if you see a list of clips that did not sync; there will be several reasons for this. Resolve will try to match up every clip you selected, and if there are any clips with missing audio or timecode (depending upon which auto-sync method you selected), then it will not sync them. Do make a note of these un-synced clips, as you can always manually sync the clips.

Additionally, if your camera did not record on-camera sound (or you forgot to turn the camera mic on) and it does not have a timecode, you can manually sync the audio with the video.

Manually syncing audio to video

This is based on a visual frame of reference for the audio, such as a hand clap within the camera frame or the use of a clapper board or even a digital slate on an iPad.

> ### Key concept – marking the take
>
> You may have seen behind-the-scenes footage of the filmmaking process where at the start of the scene a clapper board (also called a *slate* or *sticks*) is used with the name of the film, director, and so on written on it, which is then clapped to create a short sharp clapping sound. The film industry calls this *marking the take*. This clapping is important to help the audio and the video to be synched later in editing software. To get a successful sync, the clap should be clearly visible to the camera and the separate audio should be recording before the clap starts.
>
> If you do not have a clapper board, you can use your hands. It is also good to make a short sharp sound when you clap, as this is easier to see in the waveform and makes the job of manually syncing audio so much easier.

Whatever you use as a reference for the video and audio to sync, make sure that you can visually see in the video where the two objects meet to make the sound.

There is no point using your hands to make a clapping sound if you cannot see the point where your hands meet. I know this sounds obvious, but I have heard of many a frustrated editor trying to sync sound for a student film where the hand clap to sync sound was not visible or just out of the frame.

Let us see how we can manually sync sound if there is no other way to sync it:

1. Follow *steps 1 to 3*.

2. At the top right of Resolve, there is an **Embedded Audio** panel (*Figure 3.11*), which shows the audio meters by default. Select the **Waveform** tab next to the meters (*Figure 3.11*). The **Waveform** tab lets you view and scrub along the waveform of audio clips you select in the bins on the **Media** page.

3. In the **Media** page bins (named **Master**), select a video clip to sync. In the **Viewer** window, move the play-head to line up with the first visual sync point in the first clip, such as a clapper board or someone clapping.

4. Now, select in the **Media** page bins the audio clip that you want to sync to the video clip that is showing in the **Viewer**. You will see the waveform of the audio clip in the **Embedded Audio** > **Waveform** tab (*Figure 3.11*).

5. Use the waveform transport controls and scrubber bar in the Waveform Viewer to move the play-head to the audio sync point that matches the video sync point. Remember that you can use the right and left arrows to move one frame at a time in either direction to fine-tune your sync point.

    Your sync point may be a hand clap, a clapper board, or any other short, quick sound where we can see a visual reference to the sound source in the video.

    In the Waveform Viewer, the bottom half of the **Viewer** shows the entire waveform for the whole clip. The top half of the Waveform Viewer shows an enlarged section of the waveform that follows the play-head, which is useful for finding your sync point.

    So basically, you find the visual cue for the sound source in the video, such as where the hand clap meets, and then match it with the corresponding peak of the audio waveform of the separate audio clip.

6. Now that you have matched up the audio and video sync points, you just need to lock them together. Click the **Link/Unlink Audio** button (*Figure 3.11*) at the bottom right of the **Embedded Audio** panel to attach the synced audio to the video clip.

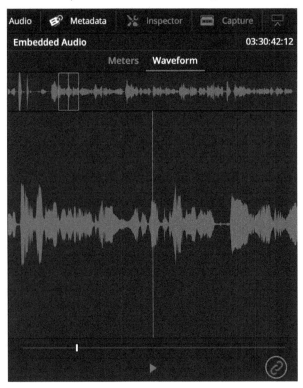

Figure 3.11 – The Link/Unlink Audio button

That's it. Your video and audio are now synced together!

Now that we have synced our separately recorded sound to the video, let us look at how we can enhance the sound further by fixing any quality issues it may have, such as that annoying electrical hum we mentioned earlier.

Fixing sound

Now, occasionally, regardless of how much effort you put into choosing the right microphone and positioning it as well as you can, there will still be issues with the sound that you have recorded. Let us look at how we can fix some of these common issues.

These audio fixing tools are **Fairlight FX** plugins built into Resolve and are found on the **Cut**, **Edit**, and **Fairlight** pages. We are going to explore the use of these tools called *plugins* from within the **Cut** page. Go to the **Cut** page, and navigate to **Effects** > **Audio** > **Fairlight FX**, where you will find the following built-in audio-fixing plugins.

De-Esser

Sometimes, when recording some people's voices, they sound sibilant. Sibilance is when someone over-emphasizes the "s" sound in words. It can be jarring after listening to this hissing sound for a long time. Thankfully, Resolve has a plugin to specifically remove this sound from dialogue or other vocals. For this exercise, record your voice saying "Suzy sells seashells by the seashore," which will create a recording containing a lot of sibilance. Import your recording into Resolve, place it on the **Timeline**, and then follow the following steps.

Let us explore how the **De-Esser** works:

1. Drag the `Fix_Sound.mp3` file that you imported at the start of the chapter onto the **Timeline**.

2. On the **Cut** page, navigate to **Effects** > **Audio** > **Fairlight FX** > **De-Esser**.

3. To apply the **De-Esser** effect, you can do one of two things:

   - Select the clip you want to apply the **De-Esser** effect to, and then double-click **De-Esser** in the **Effects** > **Audio** > **Fairlight FX** panel

   - Drag the **De-Esser** effect onto the clip to apply it

4. The **De-Esser** control panel (*Figure 3.12*) will open with a graph, showing you the part of the audio signal (frequency) that the controls will affect. To the right of the graph, there are meters that show you the **reduction** (the amount of the sibilant signal being reduced) and the **output** (the overall sound level after the effect has been applied).

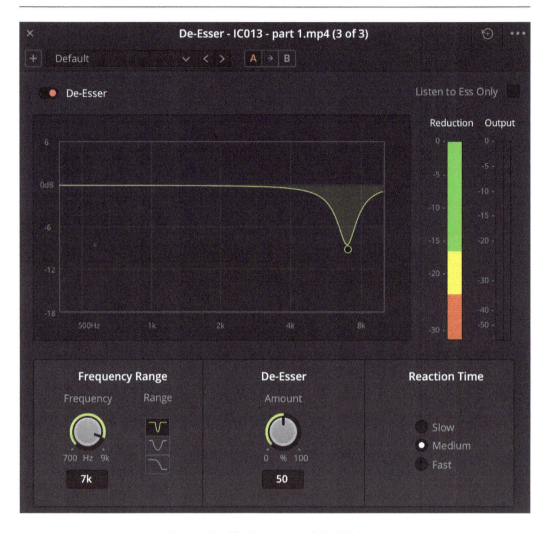

Figure 3.12 – The De-Esser Fairlight FX plugin

The following options allow you to fine-tune the targeting and reduction of the sibilance:

- **Bypass**: This is an orange toggle switch (labeled **De-Esser**) that turns the plugin on and off. Make sure this is turned on.

- **Listen to Ess Only**: This checkbox at the top right lets you listen only to the "s" sounds (esses) that are being removed. This is good to see how much of the esses are being removed and whether there are any other wanted sounds that the De-Esser is also removing erroneously.

- **Frequency Range**: These two controls let you select the frequency of the "s" sound for an individual's voice:

  - The **Frequency** knob lets you select the frequency of the "s" sound. Sibilant sounds are usually found in the frequency range of 4–8 kHz.

  - The **Range** buttons switch the mode of the De-Esser. There are three options (from top to bottom) that let you switch between **Narrow Band**, **Wide Band**, and **All High Frequency**, which reduces all audio above the selected frequency.

- **De-Esser Amount**: This knob adjusts the amount of de-essing that is applied. You will see the notch on the graph go up or down, depending upon how much de-essing you apply.

- **Reaction Time**: These radio buttons adjust how quickly de-essing is applied – in other words, how much delay before the effect is applied. There are three options here:

  - **Slow**: This slowly applies the effect

  - **Medium**: This is halfway between slow and fast

  - **Fast**: This applies the effect suddenly

Removing hum

As mentioned earlier, electrical hum can be introduced to a sound signal through a long cable run, or an unshielded or poorly earthed microphone cable. The good news is that this hum is relatively easy to identify and remove, as it exists at a specific frequency based on the electrical phase of the country that the cable is used in.

If you are based in the UK, Europe, India, China, Africa, Russia, or Australasia, your electrical frequency will be 50 Hz; if you are based in the USA, Canada, or South America, it is 60 Hz. Regardless, it is best to check what the phase of your country's electrical supply is (there are links to useful websites to do this at the end of the chapter).

Let us look at how the **De-Hummer** plugin works using the same `Fix_Sound.mp3` file that we added to the **Timeline**:

1. On the **Cut** page, navigate to **Effects > Audio > Fairlight FX > De-Hummer**.

2. To apply the **De-Hummer** effect, you can do one of two things:

   - Select the clip you want to apply the **De-Hummer** effect to, and then double-click the **De-Hummer** in the **Effects > Audio > Fairlight FX Panel**

   - Drag the **De-Hummer** effect onto the clip to apply it

3. The **De-Hummer** control panel will open and show you the following options (*Figure 3.13*):

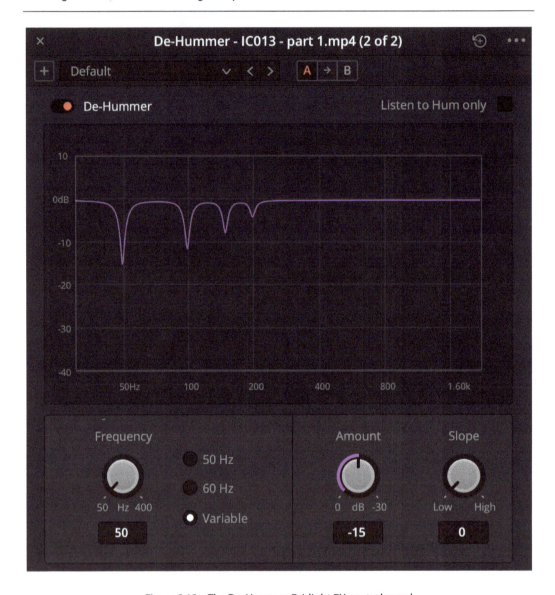

Figure 3.13 – The De-Hummer Fairlight FX control panel

- **Bypass**: This is an orange toggle switch (labeled **De-Hummer**) that turns the plugin on and off. This is called the **Bypass** switch, as the audio "bypasses" the effect when turned off, much like a car uses a motorway bypass to avoid going through a city. Make sure this is turned on.

- **Listen to Hum only**: This checkbox at the top right lets you listen only to the hum that is being removed. This is good to see how much hum is being removed and whether there are any other wanted sounds that the De-Hummer is also removing erroneously, such as dialogue.

- **Frequency**: This is the target frequency that the plugin will remove. There are radio buttons that let you select the common electrical mains frequencies of 50Hz/60Hz that most electrical hum is caused by. A knob also allows you to select other frequencies if needed.

- **Slope**: This knob adjusts the slope of the harmonics that are also removed. Harmonics are other secondary frequencies that are double the frequency of the main frequency. So, if the main frequency is 50 Hz, then the harmonics are 100 Hz; if it is 60 Hz, then they will be 120 Hz. The slope is the ratio of the main frequency to the harmonics or higher secondary frequencies that are being removed. If you set the slope to 0, then the main frequency is the priority to be removed; if the slope is set to 1, then the secondary frequencies are prioritized. A slope value of 0.5 will mean all frequencies will be targeted equally.

- **Amount**: This knob adjusts how much of the selected frequency is removed.

In addition to these controls, a graph lets you see the frequency that the plugin's controls are targeting.

Changing stereo to mono

This plugin is for fixing stereo sound recordings. As all dialogue sound is usually recorded using mono mics, you will hopefully not need this plugin. However, occasionally you may need to fix a sound clip where dialogue has been recorded using a stereo mic.

The problem with stereo sound for the human voice is that it makes the sound feel like it is coming from all around you (which is good for "the voice of God" or narration effects), but in real life, we hear the voice of a person coming from one direction, usually in the center, directly in front of us.

Thankfully, this plugin enables us to convert a clip from stereo to mono. Let us explore this useful feature, as well as briefly look at the other controls it offers. The same `Fix_Sound.mp3` clip has sound only coming out of the left speaker, as it has been incorrectly saved as a stereo clip. We need to fix this by converting the clip to mono so that the sound comes through both the left and right speakers equally:

1. On the **Cut** page, navigate to **Effects** > **Audio** > **Fairlight FX** > **Stereo Fixer**.

2. To apply the **Stereo Fixer** effect, you can do one of two things:

    - Select the clip you want to apply the **Stereo Fixer** effect to, and then double-click **Stereo Fixer** in the **Effects** > **Audio** > **Fairlight FX** panel

    - Drag the **Stereo Fixer** effect onto the clip to apply it

3. The **Stereo Fixer** control panel will open and show you the following options (*Figure 3.14*):

Figure 3.14 – The Stereo Fixer Fairlight FX

- **Format**: This is the mode that you want Resolve to use to fix the stereo. The buttons underneath (from left to right) perform the following functions:

  - **Stereo**: No fixing is done (default).

  - **Reverse Stereo**: This swaps the left and right stereo channels.

  - **Mono**: This converts the stereo clip into a mono clip by mixing the two stereo channels. This is the option we want to choose for our Fix_Sound.mp3 clip.

  - **Left Only**: The left stereo channel is sent to both the left and right stereo channels.

  - **Right Only**: The right stereo channel is sent to both the left and right stereo channels.

  - **M/S**: This is a more advanced tool that converts the stereo recording into the **Mid/Side (M/S)** format used by M/S recording microphones. A mid/side microphone is actually two mics – one mic facing forward and a stereo mic facing sideways in order to record a wider stereo sound.

- **Output Gain**: The **Left** and **Right** gain knobs allow you to apply different levels of gain on the left or right stereo outputs. This gain is applied after the **Format** mode has done its work on the stereo inputs.

There are two volume meters on either side of the plugin. The meter on the left shows the sound levels of the input of the stereo sound (i.e., the original volume levels of the clip). The meter on the right shows the sound levels of the output of the stereo sound (i.e., the new volume levels of the clip after the effect is applied).

Top tip – converting to mono

As sound volume levels are accumulative (i.e., one sound on top of another increases the overall volume of the audio), converting stereo to mono by mixing the stereo channels together will increase the overall volume of the clip. You can compensate for this by lowering the volume levels using the clip's master volume in the **Inspector** or the channel's mixer controls.

Noise reduction

Sometimes, when recording audio, we may get noise introduced by a low-quality microphone. The **Noise Reduction** plugin can reduce a range of different types of noise.

Let us explore how the **Noise Reduction** plugin works, using the same `Fix_Sound.mp3` clip already on the **Timeline**:

1. On the **Cut** page, navigate to **Effects > Audio > Fairlight FX > Noise Reduction.**

2. To apply the **Noise Reduction** effect, you can do one of two things:

    - Select the clip you want to apply the **Noise Reduction** effect to, and then double-click **Noise Reduction** in the **Effects > Audio > Fairlight FX** panel

    - Drag the **Noise Reduction** effect onto the clip to apply it

3. The **Noise Reduction** control panel (*Figure 3.15*) will open up with a graph, showing you the range of frequencies that the controls will affect, highlighted in purple.

    Two audio meters on either side of the graph show you the input sound level (on the left) compared to the output sound level (on the right). This is useful to show how much of the final audio signal is being lost due to the noise reduction being applied.

    In a drop-down menu at the top left, there are three default presets you can select, **De-Hiss**, **De-Rumble**, and **De-Rumble and Hiss**, as well as **Reset Noise Profile** to remove the presets.

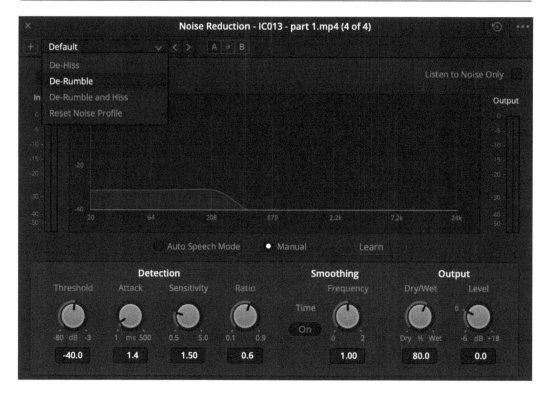

Figure 3.15 – The Noise Reduction Fairlight FX plugin

When noise reduction is being applied, the graph shows a spectral analysis of the audio being targeted, along with a purple overlay that shows what noise is being targeted.

There are two ways to reduce audio noise:

- **Auto Speech Mode**, where Resolve will try and automatically identify and reduce the noise whilst maintaining speech in the clip

- **Manual** mode, where you teach Resolve what the noise sounds like so that it only removes that sound from the clip

To enable **Manual** mode to work, you select the **Manual** radio button, click the **Learn** button, and then play back a small section of your clip that only contains the background noise that you want to remove. Resolve will listen to the sound of the noise and create a **noise profile**, which it uses to identify noise in the clip. When you have finished teaching Resolve what the noise sounds like, un-click the **Learn** button. Resolve will now use the noise profile to remove noise from the input signal of the rest of the clip.

For most people, the default settings for either the automatic or manual noise reduction methods will work perfectly fine.

However, if you want to delve a bit deeper into the controls and tweak the noise reduction results, then the following options allow you to fine-tune the targeting and reduction of the noise to your liking:

- **Bypass**: This is a orange toggle switch that turns the plugin on and off. Make sure this is turned on.

- **Listen to Noise Only**: This checkbox at the top right lets you listen only to the noise that is being removed. This is good to see how much of the noise is being removed, whether there are any other wanted sounds that the **Noise Reduction** effect is also removing erroneously, or whether too little reduction is being applied.

- **Detection**: This is where you tell Resolve what the parameters are within it to detect noise. The four knobs (left to right) underneath **Detection** (Figure 3.15) perform the following functions:

  - **Threshold** (in dB): This is the upper volume level of sound to which noise reduction is applied. Any audio below this threshold will have noise reduction applied to it. This level relates to the **signal-to-noise ratio** (**SNR**) in the original clip.

> **Key concept – SNR**
>
> SNR is the difference between the background noise level of the microphone and the upper level of the sound signal being recorded, both of which are measured in **Decibels Sound Pressure Level** (**dB SPL**), where the maximum sound level is 0 dB. For example, for a microphone that has a peak recording signal with a strength of -10 dB and a noise signal starting at and below -80 dB, the SNR will be -10 - (-80) = 70 dB.

  Audio recordings with a poor SNR will need a higher threshold value, leading to more noise reduction being applied to a clip.

  - **Attack** (in milliseconds): This is how quickly the noise reduction detection responds to variations in noise. If there is consistent noise that doesn't vary much, then a high **millisecond** (**ms**) value (longer time) of attack can be used. If there is inconsistent noise that quickly varies, then a low ms value (quicker time) of attack can be used. It is particularly useful in **Auto Speech Mode**.

> **Key concept – attack**
>
> **Attack** is how quickly any audio effect is applied to audio. The key concept here is that attack is measured in milliseconds (ms). So, a high attack value (a longer time before being applied) means the effect is applied slower, whereas a low attack value (a shorter time before being applied) means that the effect is applied quickly.

- **Sensitivity**: This knob increases the sensitivity of the noise reduction to noise. Higher sensitivity values amplify the amount of noise being detected but may also remove wanted dialogue sounds.

- **Ratio**: The **Ratio** knob controls the amount of time between the sound signal being detected and when the noise profile is being applied. A longer delay between the signal and the noise profile allows for better preservation of changes in the human voice, potentially sacrificing noise reduction.

- **Smoothing**: These options smooth the sound of the noise reduction after it has been applied to remove any unwanted sounds created as a by-product of it:

  - **Frequency Smoothing**: This smooths the signal curve across the frequencies to lessen any hard transitions as a result of the noise reduction

  - **Time Smoothing**: This toggle button enables smoothing across time as well as the frequency smoothing

  Both frequency and time smoothing blend the noise reduction into the rest of the signal. However, these may make the original audio sound a bit muddy or less defined, as they smooth out any sound transitions in the audio.

- **Output**: These options affect the output signal after the noise reduction has been applied:

  - **Dry/Wet**: This knob controls the percentage of the output mix of "wet" or noise-reduced signal to a "dry" unprocessed signal. So, a value of 0 is completely dry, whereas a value of 100% is completely wet.

  - **Makeup Gain**: This knob lets you make up (by adding gain) to compensate for any sound levels that may be lost due to the noise reduction that you have applied. This gain is applied before the dry/wet mix.

> **Top tip – resetting effects knobs to default values**
>
> It can be easy to get lost with changing the setting of so many knobs at once. Resolve lets you reset each knob to its default value by double-clicking on the middle of the knob.

Now that we have learned how to remove noise and electrical hum, and fix the stereo sound of a clip using **Fairlight FX** plugins, it is time to reveal a new additional tool in the **Inspector** that can fix all these problems using just one setting.

Voice isolation

In the latest version of DaVinci Resolve 18.1.3, there is a new tool in the **Audio** section of the **Inspector** called **Voice Isolation** (*Figure 3.16*). Voice Isolation uses **Artificial Intelligence (AI)** to analyze a clip and remove any unwanted background noise to isolate the voice. Let's try it with the Fix_Sound clip:

1. Select the Fix_Sound clip on the **Timeline** on the **Cut** page.

2. Double-click on **Voice Isolation** in the **Inspector** window to reveal the **Amount** slider (*Figure 3.16*).

3. Select the amount of voice isolation you want to apply using the slider, where **100** is voice isolation at full strength and **0** is no voice isolation applied.

Figure 3.16 – Voice Isolation

Voice Isolation will not work for every audio problem, but it is a quick and easy place to start before you need to use the more involved **Fairlight FX** plugins.

Whichever method you choose to fix audio, it is important to fix the audio first before adding any audio enhancements. Otherwise, you will be enhancing and amplifying poor-quality sound, which, of course, we do not want for our video.

So, that's it – you have now explored the range of microphone types and pickup patterns to be able to record better audio, which will save you time in Resolve.

You have also synced, fixed, and enhanced audio on the **Cut** page to get the most out of the audio you have recorded.

Summary

Here is what you have achieved in *Chapter 3*:

- You have learned why separately recorded audio is better than in-camera sound and the best practices for recording it
- Synced separate audio to video footage using waveforms
- Fixed common sound issues using **Fairlight FX**
- Used **Voice Isolation** AI to fix audio

You now know which microphones to choose to help get the best sound recording. Also, you know how to use the **Media** page to sync separately recorded audio to your video to create a cleaner sound. Additionally, you know how to work with audio on the **Cut** page to fix common audio issues without having to learn how to use **Fairlight**.

In *Chapter 4,* we will further enhance your video by adding extra audio through recording narration and voice dubbing on the **Fairlight** page, as well as adding subtitles on the **Cut** page so that your video will be more accessible across different audiences around the world.

Questions

1. True or false? It is always best to use the in-camera mic, as it is optimized for the video created by the camera.
2. True or false? You can only sync separate audio to the video if you recorded in-camera sound.
3. True or false? The **Cut** page has access to **Fairlight FX** audio plugins that enable you to fix and enhance audio.
4. True or false? **Voice Isolation** is a **Fairlight FX** plugin that isolates a voice from background noise.

Further reading

- More information about the different electricity phases for each country can be found here: https://en.wikipedia.org/wiki/Mains_electricity_by_country
- This website also has a handy downloadable PDF of each country's electrical phases: https://www.generatorsource.com/Voltages_and_Hz_by_Country.aspx

4

Adding Narration, Voice Dubbing, and Subtitles

With our short film complete, we will look at making the video accessible by adding narration, voice dubbing, and subtitles. To do this, we will start by exploring how to create and import subtitles on the **Edit** page, and then how to add narration and voice dubbing on the **Fairlight** page.

In this chapter, we're going to cover the following main topics:

- Understanding the layout of the **Edit** page
- Understanding the importance of subtitles and closed captions
- Creating and working with subtitles
- Knowing how to export and import subtitles
- Understanding the layout of the **Fairlight** page
- Knowing how to patch, arm, and record a narration or voice-over in **Fairlight**
- Using the **Automated Dialogue Replacement** (**ADR**) tools in **Fairlight** to create a voice dub

Technical requirements

For the exercise at the end of this chapter, we will walk through a practical example of how **ADR** works, where we will add spoken audio to a short clip from a Buster Keaton silent comedy called *One Week*. You can download the original film from: `https://packt.link/B5bqz`

We will also use this footage in *Chapter 6*.

Welcome to the Edit page

As easy and powerful as the **Cut** page is, it does have its limitations in order to be as streamlined as possible.

One of these limitations is that the **Cut** page cannot add subtitles. In order to be able to add subtitles, you need to use the **Edit** page. When you click on the **Edit** page, all of your video edits from the **Cut** page—including titles and effects—are instantly accessible on the **Edit** page.

Before we look at adding subtitles on the **Edit** page, let us first look at its interface to see the similarities with and differences from the **Cut** page:

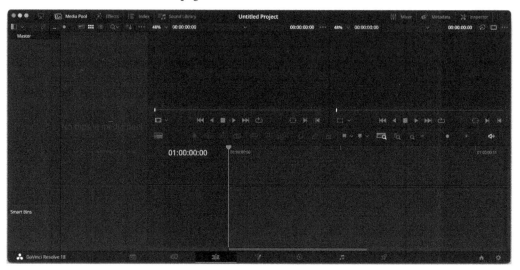

Figure 4.1: The Edit page

As you can see, the **Edit** page has the same access to the **Media Pool** and **Inspector**, which are in the same places as on the **Cut** page.

However, there are some notable differences from the rest of the interface. There is only one **Timeline** rather than the two on the **Cut** page. This **Timeline** is similar to the lower "zoomed-in" **Timeline** you have on the **Cut** page. To get a zoomed-out view like the upper **Timeline** on the **Cut** page, use the **Full Extent View** (*Figure 4.2*) button (magnifying glass over a ruler). Let's have a look at the **Timeline** buttons in more detail:

- The **Full Extent View** button is the first one at the right-hand side of the toolbar above the **Timeline**. The **Full Extent View**, just like the **Upper Timeline** on the **Cut** page, will automatically resize the **Timeline** view so that you will always see all the clips on the **Timeline** regardless of the edits you make to the **Timeline**.

- The next **Timeline View** button is the **Detail Zoom** button, which will zoom in on a small part of the **Timeline**.

- The third **Timeline View** button is the **Custom View** button, which allows you to customize the zoom level using the + and − slider to the right of it:

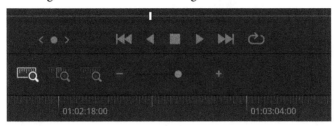

Figure 4.2: Timeline Resize and Zoom buttons (Full Extent View highlighted white)

There are two viewers rather than the one on the **Cut** page. The left-hand **Viewer** (*Figure 4.1*) is the **Media Viewer** and shows you a preview of any media that you have selected in the **Media Pool**. The right-hand **Viewer** (*Figure 4.1*) shows you a view of the media under the playhead on the **Timeline**.

Transitions and titles no longer have their own tab but are now hidden within the **Effects** tab (*Figure 4.3*):

Figure 4.3: The Edit Page Effects tab

There are also new tabs: the **Edit Index** and **Sound Library** tabs at the top left, and the **Mixer** and **Metadata** tabs at the top right.

Let us look briefly at these extra features and outline what they do.

Edit Index

This is a list of every edit on your **Timeline**. Clicking on any edit point in the list will act as a shortcut to take the playhead instantly to that edit on the **Timeline**.

Sound Library

This is a library of sound effects that you cause in your video. You can import free sound effects from Blackmagic Design and even import your own sound effect libraries. We will look deeper into this in *Chapter 5*.

Metadata

This panel shows you more information about the media clip you have selected in either the **Media Bin** or the **Timeline**.

> Key concept – metadata
>
> *Meta* is Greek for *behind/beyond*. So, essentially, metadata is the information or data hidden within a media file.
>
> This can include information recorded by the camera such as manual exposure settings, or information inputted by the camera operator such as who the director is.

Mixer

This is a simplified version of the audio **Mixer** found on the **Fairlight** page.

The **Mixer** adjusts the volume levels of the different tracks on your **Timeline**. So, you can raise the volume of a quiet track and lower the volume of a loud track to create a "mix" of the different sounds in your video. It only changes the volume of each track, so you will need to change the volume of each individual clip using the volume controls in the **Inspector** before you mix the tracks.

Now that we have a brief understanding of the **Edit** page layout, let us look at how we can create subtitles.

Importance of subtitles and closed captions

There are many reasons to create subtitles for our video. All of them relate to making sure our video is accessible to a wider audience, whether it be those who are deaf or hard of hearing or viewers whose language is different from the language spoken in the video.

Legal requirements to add subtitles to your video

Some countries have made it a law to make your video content as accessible as possible, including to those with a disability.

In the UK, under *The Public Sector Bodies (Websites and Mobile Applications) (No. 2) Accessibility Regulations 2018*, all public sector (government organizations) video content made after September 23, 2020 needs to be accessible, including web- or app-based videos. This also applies to charities that are mostly public sector-funded. Interestingly, it does not apply to the websites or apps of public sector broadcasters such as the BBC.

This is good to know, particularly if you make video content for public sector-funded organizations in the UK.

Check your own country for any legal requirements for your video to be accessible.

Subtitling and SEO

There is also another advantage of including subtitles with your video. Search engines such as Google index the text from subtitles to help improve the ranking of your video in search results. So, adding subtitles to your video is an easy way to improve your video's **search engine optimization** (**SEO**) to increase your video's audience.

For example, if your video is talking about surfing and your subtitles have closed captioning that mentions surfing, then when someone is looking for a surfing video, they have a better chance of finding yours. If you have your video subtitled into another language, then surfers from other countries will also be able to find your video.

Thankfully, Resolve makes it incredibly easy to accommodate all kinds of subtitles.

Let us look at the differences between **subtitles**, **closed captions**, and **lower thirds**.

Lower thirds

A lower third is text in the lower third of the image (hence the name) that is permanently burnt onto the video image and cannot be turned off. An example of this would be the presenter's name appearing for a few seconds when they are first introduced in the video. Some movies (for example, spy films) use lower thirds to show the translation of the language being spoken by a character when it is not their main language.

Lower thirds are applied just like any other title and are found on the **Edit** page in the **Effects** > **Toolbox** > **Titles** panel. There, you will find standard lower thirds in the **Titles** section, as well as a variety of animated **Fusion** lower thirds grouped under the **Fusion Titles** dropdown.

Subtitles versus closed captions

Unlike lower thirds, subtitles and closed captions can be turned on and off by the viewer of the video. Let's look at these in more detail:

- **Subtitles** are text at the bottom of the film that translates what is being spoken into the language of the video's audience

- **Closed captions** are in the main language of the film and are for the text description of the action of the film as well as what is being said for those who are deaf or hard of hearing

As subtitles or closed captions need to be able to be turned on or off by the viewer of the video, we need to create them in a different format from our usual titles in Resolve.

As the only difference between subtitles and closed captions is the textual information you include, when we refer to subtitles in Resolve it also applies to closed captions as well.

Let us look at how to add our own subtitles, next.

Creating and working with subtitles

On Resolve's **Edit** page, you can create subtitles and edit them in a variety of ways. Let us look at the most common methods.

Creating your own subtitles

Subtitles can only be created and edited on the **Edit** page and not the **Cut** page. However, the final result of the subtitles can be seen in the **Viewer** on all of Resolve's pages.

As with other titles, subtitles can be found in Resolve on the **Edit** page in the **Effects** > **Toolbox** > **Titles** panel. Here are the steps to creating your own subtitles:

1. On the **Edit** page, navigate to **Effects** > **Toolbox** > **Titles** (*Figure 4.4*) and scroll down to the bottom of the panel.

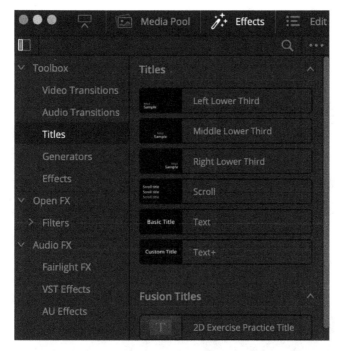

Figure 4.4: Titles panel on the Edit page

There you will see the Subtitles dropdown, under which there is only one option—Subtitle (*Figure 4.5*)

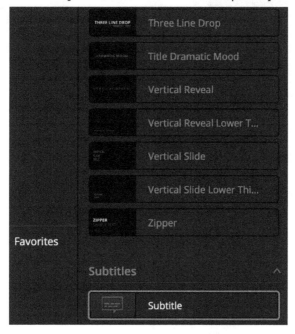

Figure 4.5: Subtitles in the Titles panel

1. Apply this **Subtitle** by dragging it onto the **Timeline**. A new **Subtitle 1** track (*Figure 4.6*) will be automatically created to accommodate the new subtitle you have just added:

Figure 4.6: Subtitle track and clip on the Timeline

Editing your subtitle

Just as with any text title, you can edit the text and format of the subtitle in the **Inspector**:

1. On the **Timeline**, select the subtitle you want to edit.
2. Open the **Inspector**. You will now see the following options in the **Inspector** (*Figure 4.7*):

Figure 4.7: Subtitle Inspector options

Let us go over what these options mean:

- **Start timecode** and **end timecode**: These are the beginning and end of the subtitle appearing on the screen based on the **Timeline**'s timecode.

  You can either manually enter new timecode numbers here or just extend or shorten the subtitle clip's duration as you would with any other clip by dragging the edges of the clip in the **Timeline** to change its length.

- **Textbox**: This is where you can change the text of the subtitle from the default text **Subtitle** to the text being spoken in the video. **8 Characters** (above and to the right of the textbox) is the number of characters in our default **Subtitle** text. This number will increase as we add text to our textbox.

- **Use Track Style**: A checkbox that changes the subtitle clip's text style between an individual style (box unticked) to a style that is applied to all the subtitle clips on the **Subtitle track** (box ticked).

- Unticking the **Use Track Style** box will reveal **Caption Style** drop-down options that allow you to change the text style of the selected subtitle clip without affecting the style of the other clips on the track.

- The **Style** tab reveals the usual text style options that we can apply to the text of all of our subtitles on the track that have the **Use Track Style** checkbox ticked. We will look at these options later in the chapter.

- **Subtitle list**: This is a list of the subtitle clips on the track, which you can navigate between using the **Prev**(ious) and **Next** buttons just above the list.

  Let us look at what the subtitle list headings mean:

- **#**: Subtitle clip number.

- **Time In/Out**: The start and end timecode for the subtitle clip—that is, its duration.

- **Caption**: The actual text of your subtitle or closed caption.

- **CPS**: This stands for **characters per second**. This is the audience's reading speed of the subtitle text in your video.

This reading speed is determined by the number of characters (individual letters) in your text and the duration of your clip. You will notice that as you type in the textbox, the CPS number will increase to accommodate the extra characters you are adding. You can lower the CPS number by stretching out the clip on the **Timeline**. You can do this by clicking on the edge of a clip and dragging it as we would any other video, audio, or title clip. Stretching or shortening a subtitle clip to fit the length of our video clip is also how we sync the clip up to fit the action in the video.

Of course, we cannot add all of our subtitle text into one clip—that would be impossible to read! We can add more subtitle clips to lower our CPS to make our subtitles easier to read. There are several ways to add a new subtitle clip, as we will see next.

Adding a new subtitle clip

Clicking the **Add New** button (*Figure 4.7*) in the **Inspector** will add a new subtitle clip starting where the playhead is placed on your **Timeline**.

This is the same as what we did earlier where we dragged the **Subtitle** effect onto the **Timeline** from the **Titles** category in the **Effects** panel.

As with many things in Resolve, there are also other ways to add a new subtitle clip:

- On the **Timeline**, right-click on the **Subtitle track** and choose **Add Subtitle** from the pop-up menu. This will add a new subtitle clip to the **Subtitle track**, starting from where your playhead is on the **Timeline**.

- If you have an empty **Subtitle track**, you can add a new subtitle clip by clicking on the **Create Caption** button in the **Inspector**. This adds a new subtitle clip to the empty **Subtitle track**, starting from where your playhead is on the **Timeline**.

The last method of adding a new subtitle clip will only work if you have selected an empty **Subtitle track** on the **Timeline**.

Let us now create a new empty **Subtitle track** on our **Timeline** to give our audience an option for viewing subtitles in an extra language.

Adding a new subtitle track

Having only one subtitle language supported in your video will limit your audience. Ideally, we will need to add a new **Subtitle track** for each additional language's subtitles.

This is easy enough to do:

1. Right-click on the header of any track in your **Timeline**.
2. Choose **Add Subtitle Track** from the pop-up menu.

A new empty **Subtitle track** will appear at the top of your **Timeline**.

So far, we have two subtitle tracks on our **Timeline**, named **Subtitle 1** and **Subtitle 2**. This does not help us know which language they each are, so let us rename them to reflect the language each one supports.

Renaming a subtitle track

It is good practice to rename the tracks in our **Timeline** to reflect their content. So far, we have not worried about changing the track names for our video and audio tracks on our **Timeline** as often, a quick edit for social media will only involve one video or audio track.

We can rename any type of track (audio, video, or subtitle) in our **Timeline** using this method:

1. Click on the **Subtitle 1** track name. This will change it into a text entry box.
2. Type in the new name of your title, such as **English CC** (**CC** being short for **closed captions**).
3. Press the *Return* or *Enter* key on your keyboard to commit your change to the track name.

Naming a **Subtitle track** is important as the name we use for the **Subtitle track** will be the name Resolve uses for our subtitle file when we export it. Some social media platforms will have particular conventions when naming a subtitle file.

Facebook naming standards

One example is that Facebook uses the `VideoFilename.[language code]_[country code].srt` naming format.

So, if I named my video *Lances-Video* and my subtitles are in British English, I would name my subtitle track *en_GB*. When exporting my .SRT subtitle file (Facebook only supports .SRT files), Resolve will add the video name *LancesVideo* to the start of the filename, so the resulting subtitle file would be called `LancesVideo.en_GB.srt`.

If you want to add other languages to your subtitles, then there is a web link to the standardized subtitle country and language codes at the end of the chapter in the *Further reading* section.

Let us look at some guidelines to help us with the font, style, and format of our subtitles.

Subtitle broadcasting standards

Looking at the broadcaster's standards for subtitles will help us create a standard look for our own subtitles that viewers are used to.

There are established broadcast guidelines as to how many characters of text we should have for each subtitle and what their CPS should be to make them readable.

Although you may not be creating your video for television or video-on-demand broadcast, reading their subtitling guidelines can be useful as a starting point for creating your own.

In the *Further reading* section, I have given links to some of the popular broadcasters' subtitles standards.

There is no international standardized subtitle format between broadcasters, so just use these guidelines as a starting point.

Resolve allows you to change the default subtitling settings to meet the standard you are aiming for.

Changing default subtitle settings

We can change the default settings for our subtitles to match the subtitle standard we are aiming for in the **Project Settings** section (*Figure 4.8*).

To go to the subtitle **Project Settings** section:

1. Click on the **Project Settings** icon (which looks like a gear wheel) in the bottom right-hand corner of Resolve (*Figure 4.8*):

Figure 4.8: Project Settings icon

2. Click on **Subtitles** in the list on the left of the **Project Settings** window (*Figure 4.9*).

3. Change your settings based on your target values for your subtitle, then click the **Save** button (*Figure 4.9*) on the bottom right of the **Project Settings** window:

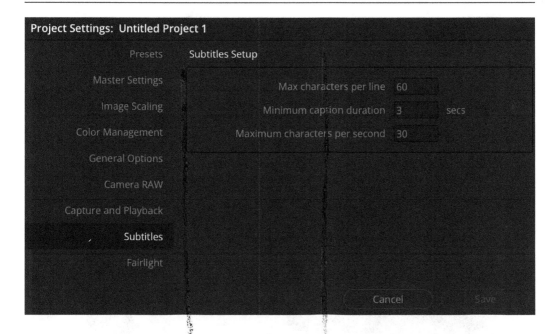

Figure 4.9: Project Settings – Subtitles

The **Subtitles** settings in **Project Settings** have the following options:

- **Max characters per line**: This is the maximum number of characters allowed on one line in a subtitle—the default is 60 characters. When you type in more characters than the line limit allows, Resolve will automatically add the new characters on a new line for the subtitle.

- **Minimum caption duration**: The default is 3 seconds. This is the minimum duration allowed for subtitles in the **Timeline**.

- **Maximum characters per second**: The default is 30 characters. This value is the maximum allowable CPS.

    Resolve automatically calculates the subtitles' CPS based on the subtitle clip's duration and character count. Resolve then shows this CPS value in the subtitle's **Inspector** rounded up or down to the nearest whole number.

    If this calculation is higher than the maximum CPS in **Project Settings**, Resolve highlights in red the value in the **Inspector** (*Figure 4.10*).

    In the following example (*Figure 4.10*), the duration of the subtitle clip is 3 seconds and the subtitle is 138 characters long, which means that the CPS value is 46. That is, 138 characters divided by 3 seconds gives us 46 CPS:

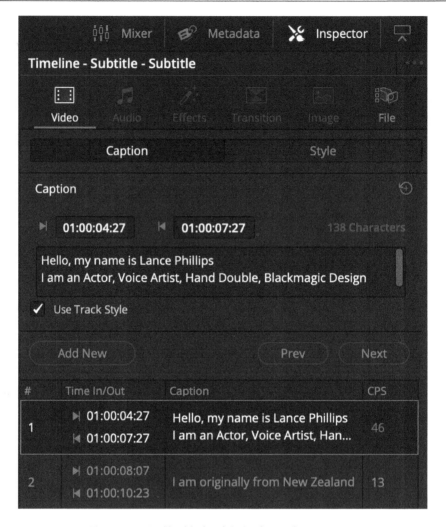

Figure 4.10: Red highlighted CPS value in the Inspector

In order to correct this red **CPS** value, I would have to either do one of the following:

- Make my subtitle clip on the **Timeline** longer.
- Reduce the number of characters in my subtitle clip. If someone is speaking too fast for the length of the clip, I would need to shorten what they are saying in the subtitles to make it readable.

Either of these options will reduce the CPS number, and the number will turn white to show that it is within the target you have set in **Project Settings**.

Let us look at how to change the subtitle font and style to meet our chosen subtitle formatting standards.

Formatting our subtitles

Resolve gives us a range of text-styling controls in the **Inspector** that will change the styling of all subtitles on the **Subtitle track**.

To do this, see the following:

1. Select a subtitle clip or **Subtitle track** (click on the track header).
2. Open the **Inspector** panel and select the **Style** tab.
3. The **Style** tab has several text formatting options, under **Style and Position**, which we have already explained the function of in *Chapter 2*.

    However, we have two additional formatting options, **Drop Shadow** and **Background**, (*Figure 4.11*) to help the text stand out from the video background. Both of these options can be toggled on or off with the red toggle switch to the left of the settings:

- **Drop Shadow**: Provides a shadow behind the text where we can change the following properties:
- **Color**: A color picker to select the shadow's color.
- **Offset**: *x*/*y* coordinates to change which direction the shadow is being cast.
- **Blur**: A slider to define how sharp the edges of the shadow are.
- **Opacity**: A slider to define how see-through the shadow is. A value of 100 is 100% solid and 0 is completely transparent.
- **Background**: Creates a rectangular box behind the text where we can change the following properties:
    - **Color**: A color picker to select the background's color.
    - **Outline Color**: A color picker to select the background's outline edge color.
    - **Outline Width**: A slider to define how thick the edges of the outline are.
    - **Corner Radius**: A slider to define how rounded the corners of the background are.
    - **Opacity**: A slider to define how see-through the background is. A value of 100 is 100% solid and 0 is completely transparent.
    - **Override Sizing**: A checkbox that turns off the automatic background resizing. In Resolve version 17.4.1 onwards, the background will automatically resize to fit just around the text. Ticking this checkbox will reveal the **Background Controls** section, where you can change the following:
        - **Width**: A slider to change the width of the background across the screen.
        - **Height**: A slider to change the height of the background across the screen.
        - **Center**: *x*/*y* coordinates to change the center position of the background.

The **Style** tab formats all subtitles on the same **Subtitle track**.

If you want to change the style of an individual subtitle without affecting the other subtitles, untick the **Use Track Style** checkbox (at the bottom of the **Caption** tab) to reveal the **Caption Style** dropdown for basic text formatting options such as **Font Family**, **Font Style**, **Color**, **Size**, **Alignment**, and **Position**. You can only override the font options using **Caption Style**, not the **Background** or **Shadow** options:

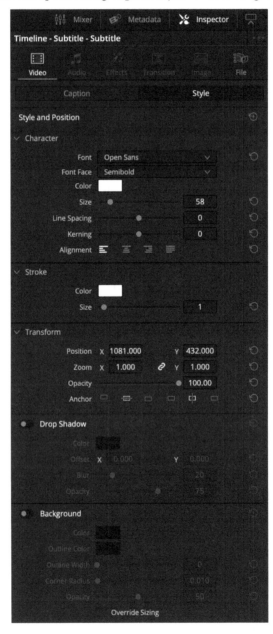

Figure 4.11: Subtitle style settings in the Inspector

Now that we have finished making and formatting our subtitles, we are ready to export them to be used on a social media platform.

Exporting and importing subtitles

Not only can you create subtitles within Resolve but you can also export your subtitles to share with someone else, or import subtitles created elsewhere to use in your project.

Exporting subtitles to upload to YouTube or Vimeo

Resolve supports two common web- and social media-friendly subtitle file formats: **SRT** (.srt) and **WebVTT** (.vtt).

SRT versus WebVTT

SRT is a basic universally recognized subtitle file format used on the internet. This subtitle file format is the preferred format for Facebook.

WebVTT is a more advanced subtitle file format that adds more functionality to your subtitles than **SRT**, such as supporting text formatting. The **WebVTT** subtitle file format is used by YouTube and Vimeo.

There are three main ways that Resolve can export our subtitles as a separate subtitle file.

Exporting subtitles in the File menu

Like most applications, Resolve allows you to export your subtitles from the **File** menu:

1. In Resolve's **Menu** bar, choose **File** > **Export** > **Subtitle…**.

2. Use your computer's **Export** dialog to choose a location to save your subtitle to and whether you want to use the **WebVTT** or **SRT** format.

   The **Without Formatting** option for **SRT** subtitles saves the subtitles without any formatting attached.

Exporting subtitles in the Subtitle track header

An even quicker way of exporting subtiles is to do it directly from the header of the **Subtitle track**:

1. Right-click on the **Subtitle track** header, and click **Export Subtitle…** from the pop-up menu.

2. As with the previous option, use your computer's **Export** dialog to choose a location to save your subtitle and whether you want to use WebVTT or SRT formats.

Exporting subtitles on the Deliver page

You can also export any subtitles you have created when you export your completed video from the **Deliver** page. Selecting the **Subtitle Settings** (*Figure 4.12*) drop-down menu at the bottom of the **Video** tab in the **Render Settings** panel will reveal more advanced options for exporting subtitles, including exporting as the broadcast-friendly IMSC1 or DFXP file formats as well as the usual WebVTT or SRT web-friendly subtitle formats:

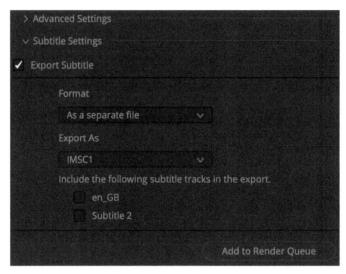

Figure 4.12: The Deliver page's Export Subtitle options

We will not go into these more advanced options here as this is a beginners' book; however, it is useful to know there are more advanced subtitle report options available in Resolve should you ever need them (should you get the need to produce subtitles for Netflix, BBC, or other broadcasters).

Of course, sometimes, to save time, we can get someone else to create the subtitles for us and then import them into Resolve.

Importing subtitles into Resolve

DaVinci Resolve at the moment only supports importing subtitle files in the **SRT** format. To import an **SRT** subtitle file, see the following:

1. In the **Edit** page, open the **Media Pool**, then right-click on either:

    *   The name of a bin in the **Bin** list

    *   Any empty space inside a bin

    In the pop-up menu that appears, choose **Import Subtitle…**.

2. Use your computer's file manager to locate the **SRT** subtitle file that you want to import (you can use the one you have just exported) and click the **Open** button.

3. The subtitle file will appear as a subtitle clip in the bin you selected. A speech bubble icon (*Figure 4.13*) on the lower left of the clip shows that it is a subtitle:

Figure 4.13: Subtitle clip in Media Pool

Positioning subtitle clips on the Timeline

Now that you have imported a subtitle clip into the **Media Pool**, you can add it to the **Timeline** as you would with any other clip:

*   Select, drag, and drop the subtitle file onto the bottom of the gray **Timeline** timecode area at the top of your **Timeline** above the video tracks, and a new **Subtitle track** will be created for your **subtitle clip**

*   If you already have a **Subtitle track** you want to put subtitles on, you can just drag the subtitle clip directly onto it

The first option can be tricky to achieve as it requires the mouse to hover over the exact right point on the **Timeline** for it to trigger creating a new **Subtitle track** automatically. Instead, it is often easier to create your own **Subtitle track** and then drop the subtitle clip onto it.

Also, if the subtitle clip does not want to drag onto the track you want it to assemble on, then lock the other **Subtitle tracks** by clicking the **Lock** icon in the track header. This will force Resolve to put the subtitle clip on the only track that is not locked.

Once the subtitle clip has been dragged onto a **Subtitle track**, it will open up and convert into a series of individual subtitle clips on the **Subtitle track**.

You can reposition the subtitle clips just like any other clip by selecting them and dragging them to line up with the other clips on the **Timeline**.

You can now reformat the subtitles you have imported or rename the **Subtitle tracks** just like you did when manually creating your own subtitles.

Now that we have added subtitles to make our video more accessible, let us look at how we can use the **Fairlight** page to add a voice dub or narration to our video to help it reach a wider audience.

But first, we will quickly look at the **Fairlight** interface so that we can find our way around it.

Welcome to the Fairlight page

Sometimes, in addition to adding subtitles, we may need to overdub our video into another language. In order to add more advanced audio techniques, we will need to visit the **Fairlight** page.

Fairlight is a fully functional **Digital Audio Workstation** (**DAW**) and is powerfully more capable than audio-editing plugins you may encounter in other video-editing software.

When you click on the **Fairlight** page, all of your audio edits from the **Cut** and **Edit** pages—including audio effects—are instantly accessible on the **Fairlight** page.

Before we look at adding narration or voice dubbing on the **Fairlight** page, let us first look at its interface to see the similarities with and differences from the **Edit** page:

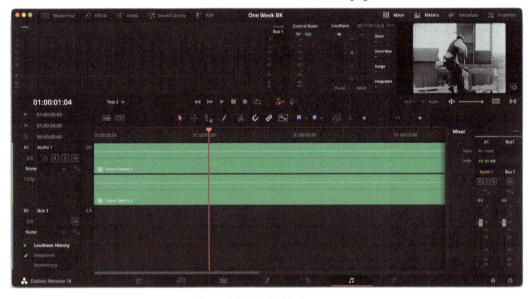

Figure 4.14: The Fairlight page

As you can see, the **Fairlight** page (*Figure 4.14*) has the same access to the **Media Pool**, **Effects**, **Sound Library**, **Mixer**, **Metadata**, and **Inspector**, which are in the same places as on the **Edit** page.

There are also some changes: **Edit Index** has been renamed **Index**, there is only one **Viewer**, and **Mixer** has more advanced controls in it. Similar to the right-hand **Viewer** on the **Edit** page, the **Fairlight** page's **Viewer** (*Figure 4.14*) shows you a view of the media under the playhead on the **Timeline**.

There are also new tabs: the **ADR** tab at the top left, and the **Meter** tab at the top right.

Let us look briefly at these extra features and outline what they do.

Index

The new **Index** tab (*Figure 4.15*) contains the **Edit Index** functionality, which, just like on the **Edit** page, is a list of every edit on your **Timeline**. Clicking on any edit point in the list will act as a shortcut to take the playhead instantly to that edit on the **Timeline**:

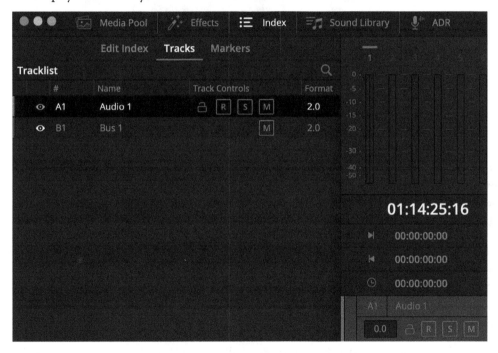

Figure 4.15: Index tab in Fairlight

However, clicking on the **Index** tab also reveals two new panels: **Tracks** and **Markers**.

Tracks

Tracks will be the default panel that you will first see when you open the **Index** tab. It contains a tracklist of every audio track in your **Timeline**. Where **Edit Index** allows you to jump between each edit point on your **Timeline**, **Tracks** allows you to jump between different tracks and either mute them (by clicking on the **M** button) or hide their visibility (by clicking on the eye-shaped icon).

Markers

This is a list of markers you have added to your **Timeline**. You can add a marker (*Figure 4.16*) anywhere on your **Timeline** (if no clip is selected) by pressing the *M* key on your keyboard once.

If you have a clip selected on your **Timeline**, pressing *M* on your keyboard will add a marker to your clip:

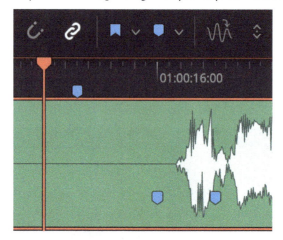

Figure 4.16: Timeline (top left) and clip markers (bottom right)

Once you have a list of markers, you can jump to each marker on the **Timeline** by double-clicking its listing in the **Markers** panel.

Meters

Meters are on by default. Clicking on the **Meters** tab will hide the audio meters and **Viewer**, which are along the top of the **Timeline**. The audio meters show the volume levels (in dB) for each of the tracks and are useful when mixing the audio levels between the tracks.

ADR

ADR is a tool that we use to replace the original video's dialogue with audio recorded afterward. This may be necessary due to poor-quality original sound needing to be replaced or, as we will explore later in the chapter, dubbing another language over the original video.

Now that we have a brief understanding of the **Fairlight** interface, let us create an empty audio track where we can store our future recordings.

Creating a new audio track

We can create any format audio track in Resolve with a number of audio channels, ranging from Mono to Dolby Atmos 9.1.6 surround sound. For voice-over or narration, we only need a new audio track to be a mono channel, as there will only be one channel of sound coming from our microphone. Here's how to do this:

Right-click on any track header. There are two options in the pop-up menu, to create a new audio track:

- **Add Track**: This will add one new audio track to the **Timeline** based on the channel type you select in the drop-down menu:

  - **Mono**: One channel of sound.

  - **Stereo**: Two channels of sound.

  - **5.1**: Five channels of surround sound with the bass frequencies sent to a separate subwoofer (the .1 in 5.1 stands for the subwoofer channel), making six audio channels in total. This is for a six-speaker surround-sound setup.

  - **7.1**: Seven channels of surround sound with the bass frequencies sent to a separate subwoofer (the .1 in 7.1 stands for the subwoofer channel), making eight audio channels in total. This is for an eight-speaker surround-sound setup.

  - **Dolby Atmos** (studio version only): Dolby Atmos-specific surround-sound setup.

  - **Adaptive**: You can select any number of channels up to a maximum of 24.

- **Add Tracks…**: In the pop-up window, you can specify how many audio tracks you want, how many channels of audio they should have, and where you want the track to appear on the **Timeline** in reference to the existing tracks (*Figure 4.17*):

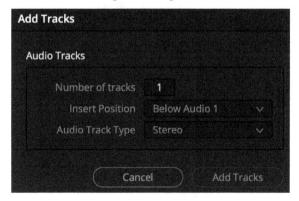

Figure 4.17: Add Tracks pop-up window

Whichever method you choose, create a new mono-channel audio track on the **Timeline** before you continue any further.

> **Key concept – tracks versus channels**
>
> Just as with the **Edit** page, the **Fairlight** page displays individual tracks of audio that can contain several clips edited together. Each track is usually used to show a new instrument or sound source.
>
> Unlike the **Edit** page, which shows several channels of sound merged into one audio waveform, the **Fairlight** page separates each track to show how many channels of sound that track has. So, a **Mono** track would only have one channel, whereas a **Stereo** track would have two channels.

Before we look at using **ADR** to create a foreign-language dub, let us look at the simpler process of adding a voice-over or narration to our video.

Sometimes, it is useful to add narration or a voice-over to our video to further illustrate the story we are trying to tell. Rather than do this in separate audio software and then import it into our edit, we can do this all within Resolve on the **Fairlight** page.

To be able to record narration in **Fairlight**, we need to patch, arm, and then record.

Patching, arming, and recording

The process of attaching a microphone to Resolve, telling it which track to record to then recording it, is called patch, arm, and record. Let us look at each one in turn ready for our voice-over recording.

Patching tracks

Before we can record from any microphone, we need to make sure it is connected to the computer, recognized by Resolve, and connected to a track, ready to record. This operation is called **patching**.

First of all, connect a microphone to your computer. If you have a built-in microphone on your computer such as on a laptop, or if you have already connected a microphone to your computer before starting Resolve, then it should be recognized in the following steps.

> **Top tip – connecting a microphone**
>
> If you are using an external microphone, then you will need to save your work and restart Resolve to force it to recognize the microphone connected to your computer.
>
> If your microphone is still not recognized, then make sure it is recognized by your computer first. If your computer does not recognize it, then restarting your computer with your microphone plugged in should fix the issue.
>
> Your computer and software apps look for connected hardware devices (such as microphones) when they are first started. Restarting Resolve (as with any software) will often solve any hardware connection issues. If this doesn't work, restarting your computer after connecting external hardware should resolve the problem. This is the case for most hardware.

Once your microphone is connected and recognized by your computer (you can check for this in your computer's sound settings), we can now patch the microphone to a track in **Fairlight**.

The term *patching* comes from the early days of audio, where to temporarily direct the audio from one source to another, an audio cable (patch cable) would be used. This temporary nature of patching (hence the word *patch*) applies in **Fairlight** to redirecting the audio from one source to a different output; it's just that we don't use cables to do it.

Let us patch our microphone into our empty track on our **Timeline**:

1. Open the **Patch Input/Output** window (*Figure 4.18*) by either doing one of the following:

   • In the **Menu** bar, choosing **Fairlight** > **Patch Input/Output…**.

   • In **Mixer**, clicking the **Input** menu (labeled **No Input**) at the top of the track you want to patch to, and choosing **Input…** from the drop-down menu. Unlike the **Menu** bar method, using the **Input** menu in **Mixer** will automatically select the track as your patch **Destination**:

Figure 4.18: Patch Input/Output window

2. Make sure that the **Source** menu is set to **Audio Inputs**, and the **Destination** menu is set to **Track Input**.

3. Just below the **Destination** menu on the right, click the audio destination that you want to patch to (empty mono track) so that it's highlighted by a white outline (*Figure 4.18*) if it's not already.

4. On the left below the **Source** menu, click the audio input (your microphone) that you want to patch from so that it's also highlighted by a white outline.

5. Click the **Patch** button at the bottom right of the window, and your microphone will be patched to your track. The patch **Input** and **Destination** buttons will now show the name of the audio channel that is patched to it on the button—that is, the **Destination** buttons will show the name of the input channel patched to it, whereas the **Input** buttons will show the name of the destination that they are patching into.

6. When you have finished patching your audio channels, click the cross in the top left to close the **Patch Input/Output** window.

Arming tracks

Before we can record audio to a track, we first need to "arm" it. Arming a track switches the track from just being able to playback audio to being able to hear the live microphone input that is patched to the track. This is useful to help us get our microphone levels right before we start recording and identify any microphone placement issues.

You will not be able to arm a track unless it already has a microphone patched to it. Once you have patched the microphone to a track, you can arm the track ready for recording and monitoring by following these steps:

1. Click the **Arm** button, labeled **R**, in either of these locations:

    *   The track header

    *   The **Mixer** channel strip

2. The button will now turn red to show that the track is now armed.

Now that the track is armed, you are now ready to record.

Recording audio

Whenever you record live audio (whether it be via a microphone, keyboard, or MIDI instrument) into Resolve, it does two things:

*   It stores it as a new audio file on your computer.

*   It puts a copy of the recorded clip in whichever bin you selected when you started recording.

Choosing where to save recorded audio clips

You can change where on your computer Resolve stores your new recording in **Project Settings**. Here's how:

1. Open **Project Settings**.
2. Open the **Capture and Playback** panel in **Project Settings**.
3. In the **Capture** section, use the **Browse** button, below the **Save clips to** textbox, to choose a new location on your computer to save your audio recording.

When you record audio into **Fairlight**, the recording begins at the position of the playhead on the **Timeline**. This is so that you can record audio into a specific position in your edit—for example, in reaction to an audio cue.

Beginning recording

Once you have placed your playhead where you want your recording to start, it is a simple case of clicking the **Record** button in the **Transport** controls above the **Timeline** to begin recording, as follows:

1. Put the playhead on the **Timeline** where you want the recording to start.
2. Click the **Record** button in the **Transport** controls above the **Timeline**. The button will turn red to show that it is recording.

A recording will start immediately, and a new audio waveform will be drawn underneath the playhead on the track you have selected.

Stopping recording

To stop recording, you can do either of the following:

*   Click the **Stop** button in the **Transport** controls above the **Timeline**
*   Press the spacebar on the computer keyboard

Recording more than one take using layers

When you record more than one take, you could decide to record them one after another on the **Timeline**; however, you can also layer the takes on top of each other at the same point on the **Timeline**.

When recording on top of a previous take on the **Timeline**, Resolve does not delete the previous recording but instead records the new take as a new layer above the previous recording, hiding the old one underneath it.

To view all of these layers and see all the previous recordings, in the **Menu** bar, select **View** > **Show Audio Track Layers** at the bottom of the drop-down menu. This option will now be ticked in the drop-down menu to show that it is active.

You will now see all the previous recordings stacked as layers on top of each other. The most recent recording will always be on top and will mute the audio of any layer underneath (*Figure 4.19*). In the following example, (*Figure 4.19*), there is one mono (1.0) track (track **A3**) with eight audio clips layered on top of each other:

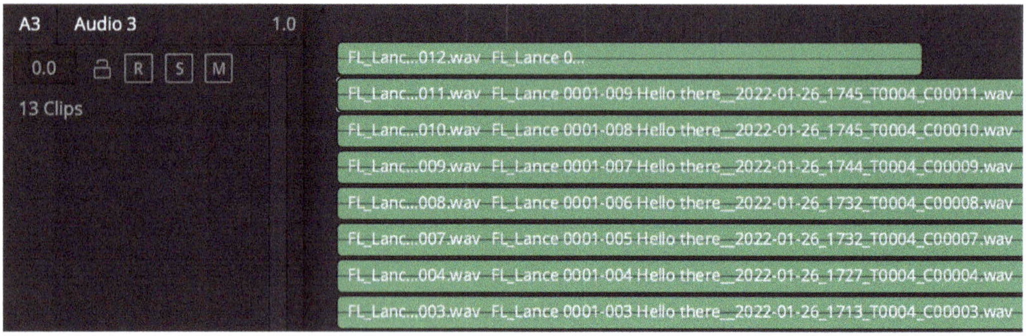

Figure 4.19: Audio Track Layers turned on

To listen to any previous recording, just drag the layer of that recording and make it the topmost layer. This is a good way to audition different audio takes or even record just parts of a voice-over without the need to start over from the beginning.

Now that you know how to record new audio directly into **Fairlight**, let us look at the **ADR** tool and its unique features that help us record replacement dialogue.

Voice dubbing in Fairlight using ADR

Automated Dialogue Replacement (**ADR**) is a particularly useful tool used in the film industry to rerecord actors' voices to get a better take or performance. The **ADR** tool can also be used to dub over your video to add dialogue in another language.

Let us walk through the different options of the **ADR** interface as if we were going to record additional dialogue in a different language.

To understand how **ADR** works, we are going to walk through a practical example where we will add spoken audio to a short clip from a Buster Keaton silent comedy called *One Week*. You can download the original film here so that you can follow along: `https://packt.link/B5bqz`.

We will also use this footage in *Chapter 6*.

Practice what you learned in *Chapter 1* by creating a new project for this exercise:

1. Create a new project and name it `Chapter_4_ADR`.
2. Import the `One_Week_512kb.mp4` file into the **Media Pool**.
3. Drag `One_Week_512kb.mp4` from the **Media Pool** onto the **Timeline**.

We are now going to add some dialogue to the breakfast scene in *One Week*. Let us select the footage we want and create a new **Timeline** with it, next.

Creating a new Timeline using selected footage

Rather than use a whole lengthy clip on the **Timeline**, which is often the case with archival footage, we can create a new **Timeline** with only part of the footage that we want to use selected from our original **Timeline**:

1. Go to the **Timeline** we just created on the **Edit** page.

2. Place your playhead on the **Timeline** at the clip that shows the **Monday 9th** calendar roughly at 01:03:18:14. Add an **In** point by typing *I* on the keyboard.

3. Place your playhead at the end after Buster's wife kisses him, roughly at 01:04:23:02. Add an **Out** point by typing *O* on the keyboard.

4. We will now create a new **Timeline** for our clips, by doing either of the following:

    - Right-clicking in the **Media Pool** and selecting **Create New Timeline Using Selected Clips and Bins…**

    - Right-clicking on the **Timeline** thumbnail in the **Media Pool** and selecting **Create New Timeline Using Selected Clips…**

5. In the resulting **Create New Timeline** dialog box (*Figure 4.20*), name the **Timeline** and make sure that **Use Selected Mark In/Out** is ticked. We have named it as **Breakfast Scene**.

6. Click on the **Create** button:

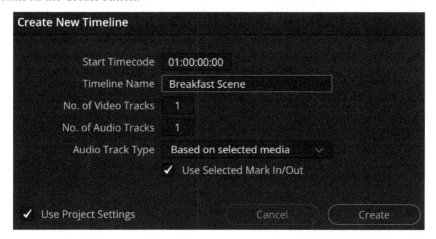

Figure 4.20: Create New Timeline dialog box

You now have a new **Timeline** in the **Media Pool** that only contains clips that have just the breakfast scene.

Before we can start to add **ADR** to the footage, we need to prepare it for the edit. As it is one single clip with no cuts between the shots, we need to divide up the clip into individual shots so that we have more control over the timings of each shot.

Adding scene cuts (Studio version only)

Rather than manually adding cuts to the footage, we use the **Detect Scene Cuts** feature in the **Studio** version of DaVinci Resolve. Here's how to do this:

1. Select the **Timeline** menu in the **Menu** bar.

2. Select **Detect Scene Cuts** in the drop-down **Timeline** menu.

    A progress bar will show (*Figure 4.21*), as Resolve looks for any sudden changes in the footage and adds cuts where the changes are:

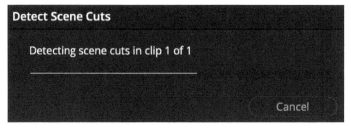

Figure 4.21: Detect Scene Cuts popup

The original clip on your **Timeline** will now be separated into individual clips for each shot.

Occasionally, Resolve will not recognize a shot change due to a dissolve between the shots. You can always add a cut manually where Resolve has not created one.

If Resolve has added a cut where there should not be one, you can remove it by following these steps:

1. Place your playhead on or just after the cut you want to remove.

2. Select the **Timeline** menu in the **Menu** bar.

3. Select **Join Clips** in the drop-down **Timeline** menu.

The two clips will now be merged into one.

There is another way to add cuts between shots. Next, we'll discuss that.

Scene Detect

If you are sent a video to edit (such as archive footage) with no access to the original clips, it can be time-consuming to manually add your own cut points to every shot to separate the scene. Thankfully, Resolve has another function called **Scene Detect** that can automatically detect individual shots and separate them by adding cuts between each shot.

The advantage of **Scene Detect** over **Detect Scene Cuts** is that with **Scene Detect**, you can change Resolve's sensitivity in being able to detect scene cuts. You can also use **Scene Detect** to manually add scene cuts that Resolve did not identify, rather than having to add the clips to the **Timeline** to cut them, as we did in the preceding method.

Let us look at how **Scene Detect** does this.

Scene Detect on the Media page

In all versions of DaVinci Resolve, including the free version, we can use **Scene Detect** on the **Media** page, as follows:

1. Open the **Media** page and select the **One_Week_512kb.mp4** video file on your hard drive in the **Media Storage** panel.

    Scene Detect will not work on clips already imported to the **Bins** section in your **Media Pool**.

2. Right-click on the **One_Week_512kb.mp4** video file and select **Scene Cut Detection…** in the pop-up menu.

    The **Scene Detect** window (*Figure 4.22*) will open with the video file you selected, ready to divide it into individual clips.

3. Click the **Auto Scene Detect** button:

Figure 4.22: Scene Detect window

4. Resolve will now play back the video and begin detecting the cuts automatically. Resolve analyzes the video, looking for sudden visual changes between frames. The **Scene Detect** window (*Figure 4.22*) will show three viewers. The first **Viewer** will show the frame before the detected cut, and the second two viewers will show the two frames of video following the detected cut. Each potential cut is placed on the **Timeline** of **Scene Detect** as a vertical line. The taller the line, the more confident Resolve will be that it has detected a cut.

Let's look at this in more detail:

- The *vertical green lines* represent each cut Resolve has detected as a potential cut.

- The *vertical white lines* are where Resolve has detected a change in the footage between frames but has not marked them as cuts.

- The *horizontal purple line* called the **confidence bar** can be adjusted up or down to change the level of confidence Resolve has in detecting cuts. Moving the bar up will make green bars shorter than the confidence bar white, eliminating them as cuts. Moving the bar down will make white bars taller than the confidence bar green, adding them as cuts.

- The numbered list on the right-hand side of the window is a list of all the cuts that Resolve has accepted. This **Cut List** functionality will show the **Scene** number, **Frame** number, and timecode (**Start TC**) for each cut. You can click on any of these listed cuts to navigate to that cut in the **Timeline**.

- The **Delete** button will remove the selected cut from **Cut List**.

- The **Add** button will add a cut where the playhead is. Sometimes, Resolve will be confused by a transition between cuts, such as a gradual fade, and not detect it as a cut. In this case, you will need to use the viewers to find and align the cut to your playhead, and then click the **Add** button to manually add the cut to **Cut List**.

- The **Slider** (at the bottom right of the window) zooms the view of the **Timeline** in or out to show the detected cuts in more detail.

5. Click on the **Add Cuts to Media Pool** button to add the cuts in **Cut List** to the **Media Pool**.

6. Close the **Scene Detect** window by clicking on the small **x** on the top left hand side of the window.

Unlike using **Detect Scene Cuts** on the **Timeline**, **Scene Detect** will add our new clips to the **Media Pool**, where we can choose which clips to add to our **Timeline**.

If we want to use **Scene Detect** rather than **Detect Scene Cuts** to detect our cuts in our breakfast scene and add them to our **Timeline**, then we will need to do the following:

1. Go to the **Edit** page.

2. Click on the **Metadata View** icon at the top of the **Media Pool** to reveal the clips' timecode:

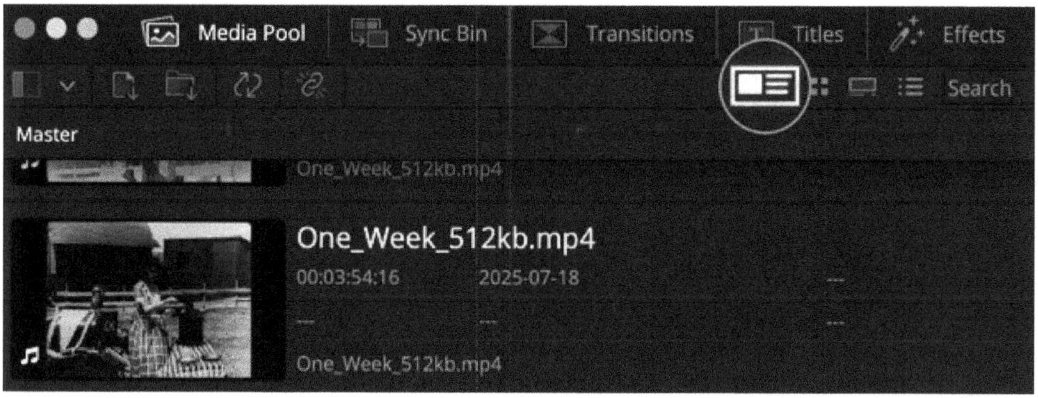

Figure 4.23: Metadata View

3. Select all clips starting from 00:03:54:16 and finishing before 00:04:23:06, by *Shift*-clicking on them.

4. Drag the selected clips onto your **Timeline**.

Let us now remove the title cards from the **Timeline** as we are replacing them with audio. There are two ways of deleting the title cards without leaving a gap on the **Timeline**, depending upon whether we are on the **Cut** or **Edit** pages.

Ripple deleting clips on the Edit page

One way of deleting clips and automatically having the rest of the clips on the **Timeline** ripple up together to close the gap left on the **Timeline** is to use the **Trim Edit** Mode:

1. Go to the **Edit** page.

2. Select **Trim Edit Mode** on the **Edit** page; this acts like the ripple mode of **Track 1** on the **Cut** page. The shortcut for **Trim Edit** is the *T* key on the keyboard. With **Trim Edit Mode** selected, any clips we delete will result in the rest of the clips rippling up to close the gap created by the deleted clip.

3. Delete the **Breakfast is ready!** and **I'll be right down!** title card clips on **Track 1**.

4. Revert back to the **Selection** mode by selecting the *A* key on your keyboard.

Ripple-deleting clips on the Cut page

An even easier and quicker way to ripple delete clips on the **Timeline** is to use **Track 1** on the **Cut** page:

1. Go to the **Cut** page.

2. Delete the **Breakfast is ready!** and **I'll be right down!** title card clips.

Now that we have ripple-deleted our title clips, let us smooth over the resulting jump cut that deleting the titles created.

Smooth Cut

We are now going to smooth over the **jump cut** left from removing the titles. Follow these steps:

1. Go to the **Cut** page.

2. Select the jump cut and click the **Smooth Cut** button.

3. Drag the handles of the **Smooth Cut** transition to make them as short as possible.

When using **Smooth Cut**, we get better results if we make the transition as short as possible, especially if there is a lot of movement in the clips on either side of the smooth cut.

> Key concept – jump cut
>
> A jump cut is where the footage within the same clip jumps forward in time, due to a section of it having been removed in the edit.
>
> Usually, we try to avoid jump cuts as they draw the audience's attention to the editing.
>
> However, they can be used stylistically, such as in horror films or music videos to create unease in the audience. The 1922 horror film *Nosferatu* used jump cuts to symbolize the unnatural movement of the vampire.

Now, let us set up **ADR** to record the missing dialogue to replace our title cards.

If you already have audio that you want to keep on **Track 1** (such as music), then first create a new audio track (**Audio 2**), as outlined earlier in this chapter, in the section called *Creating a new audio track*.

Now that you have created a new track for our ADR, let us look at the different options on the **ADR** interface.

The ADR interface

The **ADR** interface consists of three tabs: **List**, **Record**, and **Setup**. We will run through each one in the order that we will use them.

The Setup tab

This is where you set up your **ADR** session:

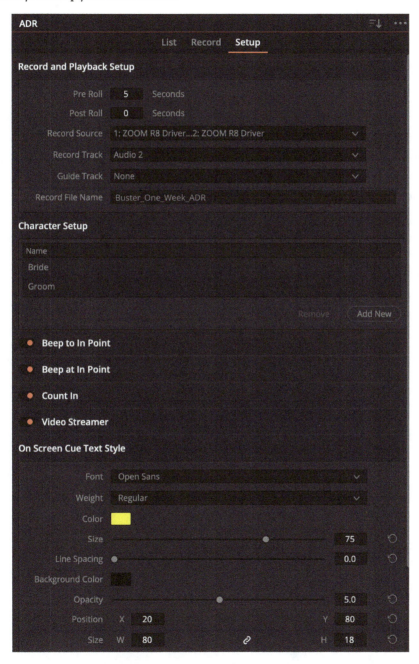

Figure 4.24: ADR interface – Setup tab

Clicking on the **Setup** tab (*Figure 4.24*) reveals a series of controls. We will focus on using the controls to get you quickly up and running with ADR.

Record and Playback Setup

This section is where we can change our settings for recording and playing back our ADR.:

- **Pre Roll Seconds**: This numerical entry box is the number of seconds that are played before the recording starts. This is useful for an actor to have some lead time before they perform.

   Set the pre-roll to 5 seconds. The default is 0 seconds.

- **Record Track**: Select the track on the **Timeline** you want to record to from a drop-down menu. In this instance, choose **Audio 2** from the drop-down menu.

- **Record Source**: This will be grayed out until you select a **Record Track** first. This is where you choose a microphone input from the drop-down menu.

   Both **Record Track** and **Record Source** work together to create a patch from your microphone input to your destination track, as well as automatically arming the track ready for recording. This means we do not need to use the **Patch Input/Output** window when patching the tracks for ADR work.

- **Record File Name**: This textbox allows you to type in a name for the audio files to be saved as so that we can easily identify it later.

- Let us name our recording `Buster_One_Week_ADR`.

Character Setup

Use the **Add New** button to add the name of each character who you will be recording replacement dialogue for. In our example, this will be `Bride` and `Groom`.

The names entered here will be available to use later in the **List** tab.

The **Remove** button will remove the name of the selected character from the list if you no longer need them or make a mistake.

Audio and visual cues

There are a series of toggle switches that enable options as visual and audio cues for the actor to begin their performance.

Once enabled, **Beeps**, **Count In**, or **Video Streamer** will provide an audible or visual countdown during the pre-roll.

Neither **Beep to In Point** or **Beep at In Point** will sound a beep without first using the **Patch Input/ Output** window to patch the **Beeps** channel from the **System Generator** source input to the **Audio Outputs** destination.

Let us instead enable **Count In** and **Video Streamer** as visual cues for our performance:

- **Count In**: A numerical counter that appears in the **Timeline** viewer that counts down to the start of the recording. Turn on this toggle switch to enable the counter.

- **Video Streamer**: Two vertical lines appear on either side of the **Timeline** viewer and move toward each other to meet in the middle of the **Viewer**. The moment they meet in the middle is the **In** point for the recording. Turn on this toggle switch to enable **Video Streamer**.

On Screen Cue text style

As the **On Screen Cue** text is white, it is hard to read over a black-and-white film. We need to change the color of the text to help it stand out from the footage behind it.

Double-clicking this header will reveal text formatting options to change the text style of the dialogue cues displayed in the **Viewer** for the actor to read on screen. Follow these steps:

1. Double-click on **On Screen Cue Text Style** to reveal the text formatting controls.
2. In the **Text Style** controls, select the color swatch to reveal your computer's color picker. Select the color yellow to help the text stand out.

Now, when you type in the character's dialogue, it will appear as yellow text over the **Viewer** for the actor to easily read.

The actor's dialogue cues are entered in the **Cue** list under the **List** tab.

The Cue list tab

Here, you can create a list of dialogue cues that the actor can read from the **Timeline** viewer. You can create a list of cues directly in the **ADR List** tab (*Figure 4.25*):

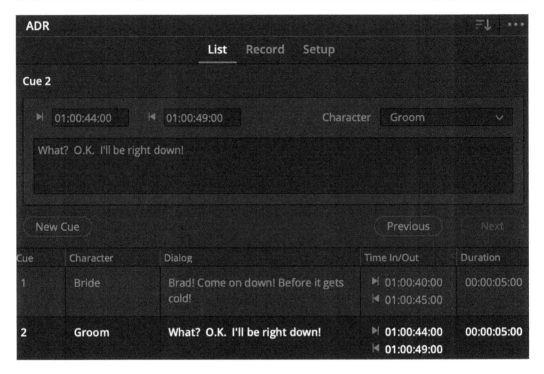

Figure 4.25: ADR cue list

Let us look at how we can create our own cue list for our Buster Keaton video:

1. Click on the **List** tab (*Figure 4.25*). You have access to the following options.

2. On the **Timeline**, line up the playhead just before the bride opens her mouth to speak at 01:00:04:00

3. **New Cue**: Click on this button to add a new cue to the list. This cue will have a **duration** of 5 seconds, using the position of the playhead on the **Timeline** as the starting timecode.

4. **Character**: This drop-down menu selects a character for the currently selected cue. The names of the characters were entered in the **Setup** tab.

   Select the **Bride** character from the **Character** drop-down menu (*Figure 4.25 this will be step 5*

5. **Dialogue entry textbox**: Click inside the box to select it. You can manually type in the bride's Brad! Come on down! Before it gets cold! dialogue directly into the textbox.

6. You can also copy and paste any text from any text editor (such as from a video script) directly into the textbox instead of typing it in manually.

- Repeat *steps 2* to *5* to add a cue for the groom, but instead of the bride, cue up the playhead for when the groom is about to speak (01:00:08:00), and then add his character name (Groom) and his dialogue: What? O.K. I'll be right down!.

The Record tab

Once you have used the **Setup** and **List** tabs to prepare the cues for recording, click on the **Record** tab (*Figure 4.26*) to reveal controls to record the cues in the **Cue** list:

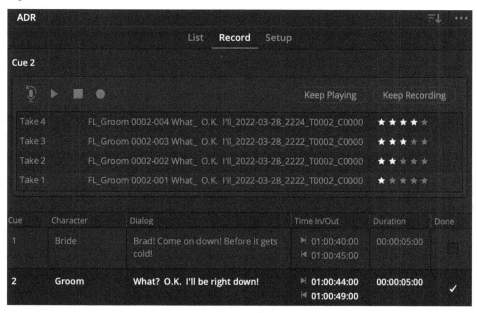

Figure 4.26: The Record tab of the ADR interface

Before we record the cue, let us rehearse the cue to get the timing without having to record any audio to the **Timeline**. You will have noticed that I have changed the words from the title cards to match the actor's lip movements but kept the contact of the scene intact.

Let us now release and then record some of our cues:

1. Select a cue in the **Cue** list.

2. Click the **Rehearse** button (microphone icon with reverse arrow).

3. **Video Streamer** and **Count In** will appear on the **Viewer** to count you in to when you should start speaking the dialogue.

4. Practice saying the text on the **Viewer** in time with the actor's lips in the video.

    Once you are happy with the timing of your rehearsal, let us now record some takes.

5. **Record**: Click the **Record** (circle) button to record the ADR for the cue. The cue will record to the selected track on the **Timeline** and will be listed as a take in the **Take** list.

You have now recorded your first ADR cue as a **take**. You can record as many takes as you like for each cue and they will be listed in the **Take** list, with the most recent one at the top.

Now, let us play back each of our takes to select the best one:

- **Play**: Clicking the **Play** (triangle) button plays the take you have selected in the **Take** list. If you haven't already selected a take, it will play the take at the top of the list.

- **Five Star Rating**: This column allows you to click on one to five stars to give each take a rating based on your initial thoughts of the take. After playing each take, give it a rating out of five stars.

- **Done** column in the **Cue** list: If you are happy with the recording of your cue for the selected cue in the **Cue** list, you can click the **Done** checkbox text, in the sixth column, at the end of the cue in the **Cue** list.

  This extra **Done** column in the **Cue** list is only available in the **Cue** list in the **Record** tab and not the **List** tab and is useful for keeping track of all the cues you have successfully recorded.

That's it! You now know how to use the **ADR** tools to overdub other languages over your original audio recording.

Of course, it is beyond the scope of this book to teach you any foreign languages, so you will need to find your own voice-over artists fluent in the language you require for your video's audience.

Summary

Here is what you have achieved in *Chapter 4*:

- Discovered the differences between the **Cut** and **Edit** page layouts

- Added different format subtitles to your video and edited them

- Discovered the unique layout of the **Fairlight** page

- Patched, armed, and recorded a narration or voice-over in **Fairlight**

- Used the **ADR** tools in **Fairlight** to create a voice dub for a video

In *Chapter 5*, we will look at creating a sound library and importing sound effects into **Fairlight**. We will also look at how to audition sound effects and how to apply them to a video in the **Timeline**. We will then change their speed and timing to fit the video.

Questions

1. True or false? All of your edits from the **Cut** page are instantly accessible on the **Edit** page.

2. True or false? Resolve only supports SRT subtitle files, so this is the only file format you can export your subtitles in.

3. True or false? **Fairlight** is a video-specific audio plugin. If you want to do a full audio edit, you will need to export your audio into an external DAW.

4. True or false? Pressing the *M* key on your computer keyboard will mute an audio track in **Fairlight**.

5. True or false? In order to record from a microphone into **Fairlight,** you will need to patch the microphone, arm the track, then record.

6. True or false? **ADR** stands for **Automatic Dialogue Recording**.

Further reading

More information about subtitles' broadcast standards:

BBC:

- `https://www.bbc.co.uk/accessibility/forproducts/guides/subtitles/`
- `https://www.bbc.co.uk/academy-guides/how-do-i-create-subtitles`

Netflix:

- `https://partnerhelp.netflixstudios.com/hc/en-us/categories/1500000000781-Timed-Text-Resources`
- `https://partnerhelp.netflixstudios.com/hc/en-us/articles/217350977-English-Timed-Text-Style-Guide`

Amazon:

- `https://videodirect.amazon.com/home/help?topicId=G201979140`

Channel 4:

- `https://www.channel4.com/media/documents/corporate/foi-docs/SG_FLP.pdf`

Subtitle file naming standards:

- `https://loc.gov/standards/iso639-2/php/code_list.php`

Facebook subtitle naming standards:

- `https://www.facebook.com/help/1528795707381162`

SRT file info:

- `https://en.wikipedia.org/wiki/SubRip#SubRip_text_file_format`

WebVTT file info:

- `https://w3c.github.io/webvtt/`

YouTube instructions on adding subtitles:

- `https://support.google.com/youtube/answer/2734796?hl=en-GB&ref_ topic=7296214#zippy=%2Cupload-a-file`

Vimeo instructions on adding subtitles:

- `https://vimeo.zendesk.com/hc/en-us/articles/224968828-Captions- and-subtitles#h_01FPAZ98X0B9ESFP23KMXGVTYB`

Accessibility requirements for UK public sector videos:

- `https://www.gov.uk/guidance/accessibility-requirements-for- public-sector-websites-and-apps`
- `https://www.legislation.gov.uk/uksi/2018/852/contents/made`

Archive videos to practice with:

- `https://archive.org/details/OneWeek`

5
Creating Additional Sound

In this chapter, we will look at enhancing the vocals you created in *Chapter 4*, using **Fairlight FX** on the **Cut** page. We will also explore how to import third-party audio FX plugins to add more functionality. Additionally, we will create a **sound effects** (**SFX**) library into which we can import **SFX** into **Fairlight**. We will then look at how to audition those SFX and how to apply them to the video in the **Timeline**.

In this chapter, we're going to cover the following main topics:

- Enhancing the vocals using **Fairlight FX**
- Importing audio effects plugins
- Importing an SFX library into **Fairlight**
- Auditioning and adding SFX

Technical requirements

For the exercises in this chapter, we will be continuing to work on the audio project we created in *Chapter 4*. If you have not completed *Chapter 4*, then I suggest you quickly follow the exercises there so that you can get the most out of this chapter.

Enhancing the vocals

In the previous chapter, you learned how to use the tools in **Fairlight** to record your voice directly into Resolve from a microphone connected to your computer. Now that we have recorded our voice-overs or ADR directly into Resolve, we can enhance the sound quality of the vocals by lifting the frequencies that we want people to pay attention to and lowering other unwanted frequencies into the background; this is called **sweetening**.

Here, we are going to look at three **Fairlight FX** audio plugins that enhance the vocals, as our listeners will often switch off if they cannot hear what is being said or it is difficult to listen to. Like the audio fixing plugins, all of these plugins can be found on the **Edit**, **Fairlight**, and **Cut** pages. For our examples, we are going to use the **Cut** page. On the **Cut** page, navigate to **Effects** > **Audio** > **Fairlight FX**, and then to the specific plugin. Let's take an in-depth look at each of these vocal-enhancing plugins, starting with the **Pitch** plugin.

Pitch

This audio plugin changes the audio pitch without changing the clip speed. It is great to add slightly more bass to a person's voice to add a bit more authority.

Figure 5.1: The Pitch Fairlight FX

Let us explore how the **Pitch** plugin works:

1. On the **Cut** page, navigate to **Effects** > **Audio** > **Fairlight FX** > **Pitch**.

2. To apply the **Pitch** effect, you can do either of the following:

    *   Select the clip you want to apply the **Pitch** effect to, then double-click the effect in the **Effects** > **Audio** > **Fairlight FX** panel.

    *   Drag it onto the clip to apply it.

3. The **Pitch** control panel (*Figure 5.1*) opens with the following controls:

    *   **Bypass**: An orange toggle switch at the top-left of the window, which turns the plugin on and off. Make sure this is turned on.

- **Pitch**: These knobs control the selection and direction of the pitch change:

  - **Semitones**: This knob changes the pitch by whole semitones. Negative values lower the pitch, and positive values increase the pitch.

  - **Cents**: This knob is for more subtle pitch changes. It changes the pitch up or down in increments of 1/100th of a semitone. This is great if you want the effect to be more subtle and less noticeable.

- **Mid/Side**: A toggle switch turns this effect on (it is off by default) and a knob balances the level of pitch correction for **Mid/Side** recordings between the mid or side microphones. Strictly for **Mid/Side** recordings, so the switch can be left in the *off* position for most recordings.

- **Output**: This knob affects the output signal after the pitch has been applied:

  - **Dry/Wet**: This knob controls the percentage of the output mix of the "wet" or noise-reduced signal to the "dry" or unprocessed signal. So, a value of **0** is completely dry, whereas a value of **100%** is completely wet. Mixing in some of the dry/unaltered voice will add more depth and naturalness to the change.

Next, we will talk about another popular vocal-enhancing plugin: **Vocal Channel**.

Vocal Channel

This plugin is for all-purpose changes to the human voice, whether it be singing or the spoken word. The controls for **High Pass Filtering**, **Equalizer**, and **Compressor** are combined within this one plugin.

Let us explore how the **Vocal Channel** plugin works:

1. On the **Cut** page, navigate to **Effects** > **Audio** > **Fairlight FX** > **Vocal Channel**.

2. To apply the **Vocal Channel** effect, you can do either of the following:

   - Select the clip you want to apply the **Vocal Channel** effect to, then double-click **Vocal Channel** in the **Effects** > **Audio** > **Fairlight FX** panel.

   - Drag it onto the clip to apply it.

3. The **Vocal Channel** panel (*Figure 5.2*) opens with the following graphs. The graph on the left lets you see the frequency that the **High Pass** and **Equalizer** controls are targeting. The **Dynamics** graph (called this as it shows the **dynamic range** of the audio) on the right shows you the effects of the **Compressor** controls. An audio meter to the right of the graphs will show you the **Output** sound level to enable you to monitor the final output of all the controls.

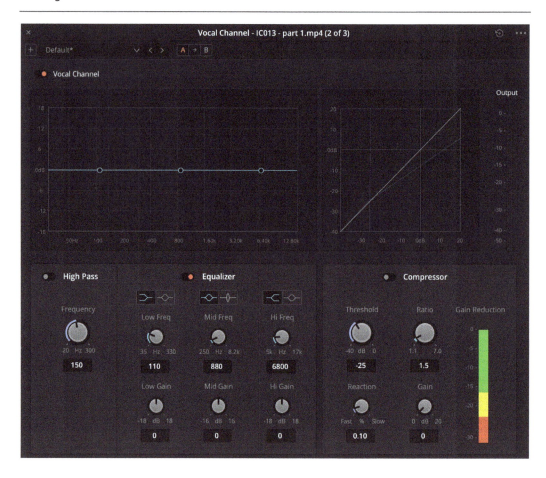

Figure 5.2: The Vocal Channel Fairlight FX

Let us explore the **Vocal Channel** controls in more detail:

- **High Pass**: This is a filter that reduces the lower-end bass frequencies. A toggle switch turns it on (it is off by default), and a knob sets the frequency level below which the bass reduction is applied. It is called a **High Pass** filter as it allows higher frequencies through but stops lower-end frequencies. It is effective to remove the "boominess" from some mics created when the singing or speaking talent gets too close to the mic.

- **Equalizer**: A three-band graphic equalizer (toggle enabled by default) for adjusting specific frequencies for speech. The controls are split for **Low**, **Mid**, and **Hi Freq** (frequency) and **Gain** adjustments:

  - **Low/Mid/Hi** selection mode: These selection buttons change the shape of the frequency selection based on the selection curve shown on the button (*Figure 5.2*).

- For example, the **Mid** range selection buttons allow you to broaden or narrow the frequencies included on either side of the core frequency selected.

- The **Low** and **Hi** range selection buttons allow you to toggle between a threshold selection (where all frequencies above for **Hi**, or below for **Low** the selection are affected) or a more targeted frequency selection:

  - **Low/Mid/Hi Freq** (Hz): These knobs allow you to select the core frequency you want to change the gain for.

  - **Low/Mid/Hi Gain** (dB): These knobs allow you to raise or lower the level of the selected frequencies.

- **Compressor**: This (toggled off by default) takes the loud sounds and compresses them down to be closer to the quieter sounds. This is good if a person's speech volume levels often alter between high and low, and makes it easier to listen to. Common examples of the use of compression are in radio commercials or film trailers where the sound of the speaker's voice stays steady in volume throughout. **Compressor** has the following controls to select and adjust the compression effect:

  - **Threshold** (dB): The **Threshold** knob sets the frequency level above which compression occurs.

  - **Reaction**: The **Reaction** knob adjusts how fast the compression is applied when the audio frequency goes above the threshold. **Fast** quickly applies compression, whereas **Slow** takes longer before applying compression.

  - **Ratio**: This knob adjusts the compression ratio, which is the amount of compression applied above the threshold set.

    The first number in the ratio is the strength of the input signal before compression, whereas the second number is the strength of the output signal after compression.

    So, a compression ratio of 4:1 means that for every 4 dB of sound input, the compressor allows 1 dB of sound output. So, a 1:1 compression ratio means no compression is being applied (for every 1 dB, 1 goes through), whereas the max setting of 7:1 is heavily compressed.

  - **Gain** (dB): The **Gain** knob compensates for any audio levels lost due to the compression by raising the gain of the output audio levels. This effectively raises the volume level of all the sounds in the clip after lowering the lowest sounds using compression. The result is that the average sound level is higher.

Dialogue Processor

Dialogue Processor contains a mixture of audio-fixing and enhancing tools associated with speech.

The **De-Rumble**, **De-Pop**, **De-Ess**, **Compressor**, **Expander**, and **Excite** controls each have simplified controls that focus on common adjustments needed for dialogue.

Let us explore how the **Dialogue Processor** works:

1. On the **Cut** page, navigate to **Effects** > **Audio** > **Fairlight FX** > **Dialogue Processor**.

2. To apply **Dialogue Processor**, you can do either of the following:

   - Select the clip you want to apply **Dialogue Processor** to, then double-click **Dialogue Processor** in the **Effects** > **Audio** > **Fairlight FX** panel.

   - Drag it onto the clip to apply it.

3. The **Dialogue Processor** panel (*Figure 5.3*) opens with two audio meters on either side of the controls to show you the **Input** sound level (to the left) compared to the **Output** sound level (to the right). This is useful to show how much of the final audio signal is being lost due to the **Dialogue Processor** controls being applied.

Figure 5.3: The Dialog Processor plugin

Dialogue Processor shows the following controls, each with their own individual toggle switch to enable or disable them:

   - **De-Rumble**: The **Frequency** knob selects the audio frequency that the **De-Rumble**, as the name suggests, will target to reduce the bass tones of the input signal. Good for reducing the rumble from mic handling or the lav/lapel mic rubbing on clothes.

- **De-Pop**: This tool, as the name suggests, will reduce the "popping" sounds created by air blowing across the mic when somebody speaks into it. This is good for when someone particularly emphasizes the "p" and "b" sounds in their speech. It has two knobs:

  - **Frequency**: This knob selects the audio frequency that **De-Pop** will target to reduce the popping sound.

  - **Amount**: This knob changes the amount of the de-popping applied to the target frequency. A dB-level meter to the right of the knob shows the amount of dB reduction being applied. The meter is color-coded to help you set your target reduction: green is good, whereas the red levels of -25 dB and below are excessive.

- **De-Ess**: This is essentially the same as the **De-Esser** effect covered in the *Fixing sound* section in *Chapter 3*.

- **Compressor**: This is a simplified version of the **Compressor** in the **Vocal Channel** covered in the previous section about the Vocal Channel. The **Fast** and **Slow** buttons select the speed of attack.

- **Expander**: This is the opposite of **Compressor**. Where **Compressor** reduces the dynamic range (the difference between the loud and quiet sounds) of the audio, **Expander** increases the dynamic range of the audio. So, in effect, **Expander** makes quiet sounds quieter and loud sounds louder. **Expander** has the following controls:

  - **Threshold**: This knob selects the audio frequency around which **Expander** will target to expand the dynamic range of the audio. It turns down the volume when the signal level falls below the threshold and turns up the volume when the signal level goes above the threshold.

  - **Range**: This knob changes the extra dynamic range that **Expander** applies around the target threshold. So, a value of **0** dB means that the dynamic range has not changed, whereas a value of **30** dB means that the dynamic range between the quietest and loudest sounds is increased by 30 db. The meter is color-coded to help you select your target expansion range: green is good, whereas the red levels of -25 dB and below are excessive.

  - **Fast** and **Slow** buttons: Select the speed of the attack.

- **Excite**: This effect adds more treble to the audio to add more vocal interest if the speaker's voice is monotone. It has the following controls:

  - **Amount**: This knob selects the amount of **Excite** applied to the audio.

  - **Female/Male**: These buttons select which target frequencies to excite depending on whether the target voice is male or female.

We have now covered a good range of the **Fairlight FX** audio plugins built into Resolve. However, Resolve does not limit you to just using its own plugins; it also has the feature of importing third-party plugins to add extra functionality.

Importing audio effects plug-ins

One advantage of using Resolve is that, like any other **Digital Audio Workstation (DAW)**, it has the ability to import extra audio plug-ins in order to expand its audio capabilities.

Resolve supports two common audio plug-in formats, AU and VST:

- **AU plug-ins**

  The **Audio Unit (AU)** plug-in was created by Apple. So, if you are using an Apple Mac, you will find that Resolve automatically recognizes the AU plug-ins already installed on your Mac.

- **VST plug-ins**

  The **Virtual Studio Technology (VST)** plug-in is a standard audio effect plug-in created by Steinberg.

  VST plug-ins come as either effects plug-ins or instrument (VSTi) plug-ins, where you can play a virtual instrument or synthesizer.

> **Top tip – finding VST and AU effects**
>
> There are plenty of independent third-party plug-in developers on the internet who offer both free and paid VST and AU plug-ins you can download. Just do a quick web search to discover pages of results.
>
> Make sure you try out their free samples first before purchasing one to ensure that they work with your version of Resolve.

Let us look at how we can import extra VST or AU plug-ins into Resolve.

For the following section on installing a plug-in, you can either download your own plug-ins to install or install this free *Protoverb* plug-in I found here: `https://u-he.com/products/`. This Protoverb plug-in has presets that simulate the reverb of different room sounds, as well as the ability to create your own room reverb presets.

Installing AU or VST plug-ins into Resolve

The following example uses an Apple Mac interface; however, the steps for a PC are very similar and the same as installing any other software you have already installed on your computer. If you are unsure about how to install software on your computer, please consult your operating system's (OS) help file or user manual:

1. In the **Finder** (*Figure 5.4*), navigate to the `.pkg` file for the plug-in you just downloaded.

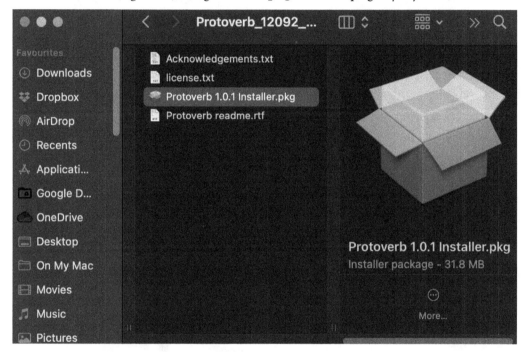

Figure 5.4: Package installer

2. Double-click the `.pkg` file (*Figure 5.4*) to open the installer.
3. Follow the instructions in the package installer (*Figure 5.5*) to install it on your Mac or PC.

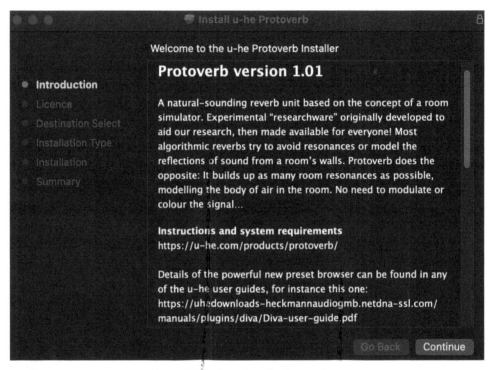

Figure 5.5: Package installer instructions

4.   Select the installation options for the plug-in you want to install (*Figure 5.6*) and click **Continue** to progress through the pages for the rest of the installation.

Figure 5.6: Installation options

5. The plug-in is now installed on your computer.

6. Close DaVinci Resolve and re-open it to make it search for any new plug-ins installed on your computer. If this does not work, you may need to restart your computer so that your computer can recognize any new plug-ins you have just installed.

Now let us check to see whether your plug-in is installed correctly:

1. On the **Cut** page, navigate to **Effects** > **Audio** and choose either **AU Effects** or **VST Effects** (*Figure 5.7*).

2. You should now see your plug-in installed and ready to be applied.

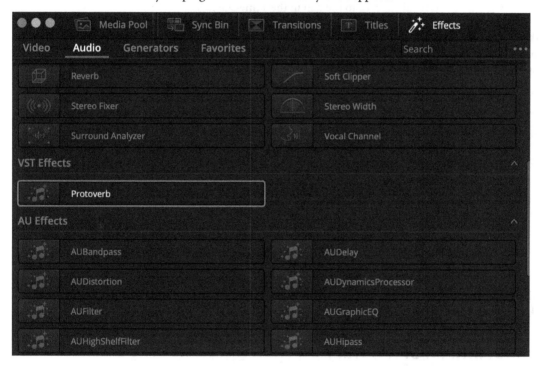

Figure 5.7: Installed plug-in

Now that we have imported some plug-ins, let us apply one to our footage.

Applying plug-ins

Applying a VST or AU plug-in to your audio is exactly the same as applying a **Fairlight FX** plug-in:

1. On the **Cut** page, navigate to **Effects** > **Audio** and choose either **AU Effects** or **VST Effects** (*Figure 5.7*).

2. Choose the plug-in you want to apply and then you can do either of the following:

   - Select the clip you want to apply the effect to, then double-click the effect in **Effects** > **Audio** > **AU/VST Effects**.

   - Drag the selected effect onto the clip in the **Timeline** to apply it.

Importing plug-ins that can fix or add effects to audio is great, but sometimes we want to add short audio clips, called **SFX** or Foley, to enhance the ambiance of the story we are telling, such as the sound of a door slamming, footsteps, or the background sound of a busy cafe.

Key concept – Foley

SFX that are created artificially, especially for a film, are called Foley, compared to just recording a real sound. For example, recording the sound of gloves flapping to represent the sound of bird wings flapping is Foley. Foley, in this example, would be used instead of the hassle of trying to record a bird's wings in flight.

Foley is often used when recording the actual sound would be technically impossible (e.g., bird wings flapping) or has a fantasy element (such as laser fire).

Foley effects are named after Jack Foley, who pioneered their use for Universal Pictures in the early days of audio cinema in the 1920s. A person who creates Foley sounds is called a Foley artist.

The good news is that we can import an **SFX library** directly into Resolve to use in **Fairlight** for this purpose.

Importing an SFX library

One of the great things about the **Fairlight** page in Resolve is that it allows you to add SFX from within Resolve without having to clutter your media bin.

Blackmagic Design helpfully provides over 500 royalty-free SFX for you to use in your projects, even commercial ones.

Let us download these free SFX to use in future exercises.

Importing the Fairlight Sound Library

Here are the links from where you can download the SFX, depending on your computer's **OS**:

- *Mac*: `https://www.blackmagicdesign.com/uk/support/download/05acbc36dbba4519a4972b7ddce31810/Mac%20OS%20X`

- *PC*: `https://www.blackmagicdesign.com/uk/support/download/05acbc36dbba4519a4972b7ddce31810/Windows`

- *Linux*: `https://www.blackmagicdesign.com/uk/support/download/05acbc36dbba4519a4972b7ddce31810/Linux`

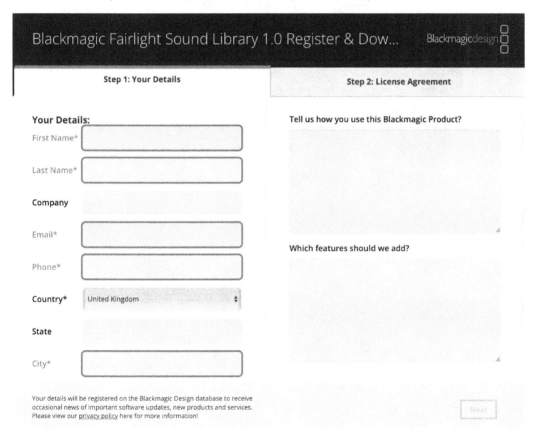

Figure 5.8: Blackmagic Fairlight Sound Library download

Once you have registered your details and agreed to the *Sound Effects License Agreement* (*Figure 5.8*), and downloaded the free SFX, you can now import them into Resolve:

1. You will now have a Software Installer file on your computer called `Blackmagic_Fairlight_Sound_Library_Mac`. Double-click this to open the installer (*Figure 5.9*).

2. Double-click the **Install Fairlight Sound Library 1.0** icon (*Figure 5.9*).

Install Fairlight Sound
Library 1.0

Figure 5.9: Fairlight Sound Library installer

3. Follow the instructions on each page of the installer (*Figure 5.10*).

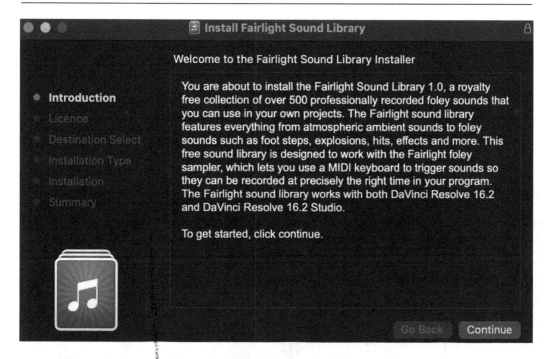

Figure 5.10: Sound Library installer pages

You now have over 500 Foley SFX installed in Resolve that you can use in your projects.

However, what do you do if you want to import your own SFX that you have found elsewhere? How do you get individual Foley sounds into Resolve without an installer?

Let us answer those questions now.

Creating your own SFX library

Creating your own SFX library is a two-stage process.

First, you need to create a Resolve database within which to store your SFX library; then, you can create an SFX library within that database.

One advantage of creating a database for your SFX, instead of just importing sounds into the **Media Pool**, is that it is indexable (hence, searchable) and you can share it across different projects and computers.

Creating a new database for SFX

As of Resolve version 18.1, databases are now called project libraries. So, if you are using an older version of Resolve, substitute the term **Project Library** for database as they are essentially the same thing.

It is good practice to have a separate dedicated project library for your SFX rather than using your current project's database. This prevents your project library from becoming too large. It also means you can easily identify where all your SFX are rather than trying to remember which project you imported them into.zz

Let us create a dedicated SFX project library now:

1. Go to **Project Manager** (the house icon at the bottom right).

2. Open the **Project Libraries** sidebar by clicking the **Show/Hide Project Libraries** icon (*Figure 5.11*) at the top left of the **Projects** window.

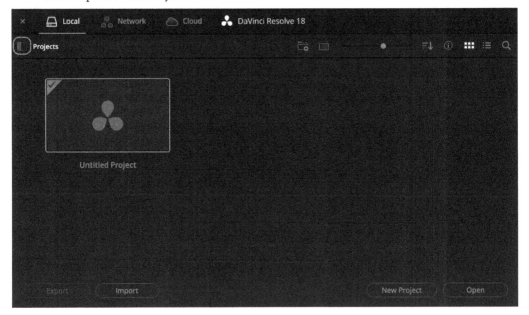

Figure 5.11: The Show/Hide Project Libraries icon

3. Click the **Add Project Library** button at the bottom of the **Project Libraries** sidebar (*Figure 5.12*).

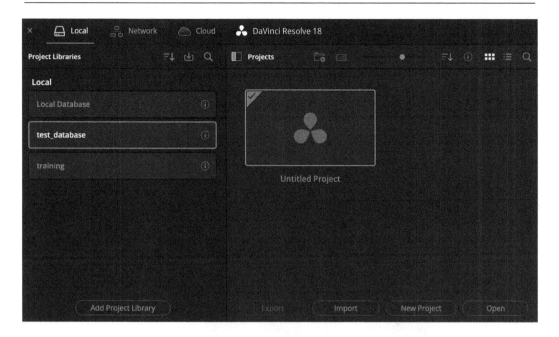

Figure 5.12: The Project Libraries sidebar

4. An **Add Project Library** window will appear (*Figure 5.13*). In the **Add Project Library** window, click the **Create** tab, if it is not already selected.

Figure 5.13: The Add Project Library window

5. In the **Add Project Library** window, complete the following information:

    • **Name**: Type the name of your project library (e.g., my_sound_fx).

    • **Location**: This is the location on your computer where you are going to save your project library:

        i.   Click on the **Browse** button to open your computer's file manager.

        ii. Navigate to where on your computer you want to store your project library, create a new empty folder, and give it a recognizable name (e.g., Resolve Sound FX Db). Resolve will only store a new project library in an empty folder.

        iii. Select the new project library folder you have just created, and if prompted by your computer's file manager, open it.

    • **Create**: At the bottom of the **Add Project Library** window, click the orange highlighted **Create** button (*Figure 5.14*).

Figure 5.14: The completed Add Project Library window

A new project library called my_sound_fx (if you used my example, like in *Figure 5.14*) will appear in the **Project Libraries** sidebar under the **Local** section.

6. In the **Project Libraries** sidebar of **Project Manager**, select the project library where the project you are working on is stored. If you haven't created any new project libraries before now, your projects will be stored in the default local database called **Local Database**.

7. Close the **Project Libraries** sidebar by clicking the Show/Hide Project Libraries icon (*Figure 5.11*) at the top left of the **Projects** window.

> **Top tip – naming project libraries**
>
> Project library names can only accept lowercase characters. To make project library names more understandable, you can use an underscore (_) to separate words in the name.
>
> There is no need for the project library and the folder that contains the project library to have the same name. Folder names don't have the same naming restrictions as a project library. Therefore, they can be given easier-to-understand names.
>
> One advantage of having a separate project library for the SFX is that you can name it based on where you sourced your SFX from. This can help when tracking down licensing agreements for the SFX you use in your project (e.g., `freesfx` or `bbc_licensed_sfx`).

Creating new project libraries in Resolve can also be useful for keeping your different projects or client work separate from each other.

Using separate project libraries for each client makes **Project Manager** less cluttered, and your projects become easier to find rather than scrolling through other clients' work.

Also, having many smaller project libraries results in Resolve being able to operate faster, as it doesn't have to search a large database to find your project.

We are now ready to create our own Sound Library.

Creating a new Sound Library

Now that we have a project library for our SFX, let's add some sounds to it.

First, you will need to have some new SFX in a folder already downloaded onto your computer, ready to import into **Fairlight**.

If you don't have any new SFX, a simple web search will reveal a multitude of royalty-free SFX that you can try out. Try web searching for `Free SFX` or `Free Foley`.

Or you can download some SFX from the links in the *Further reading* section at the end of this chapter.

Once you have downloaded the SFX onto your computer, let us import them into **Fairlight**:

1. Go to the **Fairlight** page.
2. Click the **Sound Library** tab to open the Sound Library (*Figure 5.15*).
3. Click the **Database** button at the top-right of the **Sound Library** panel to reveal a drop-down menu.

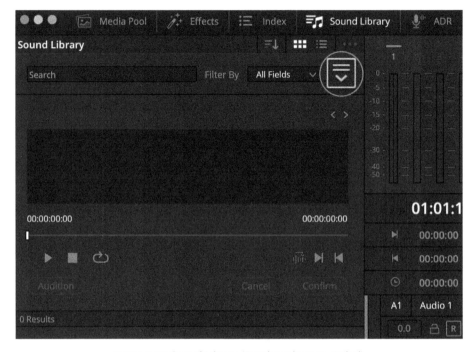

Figure 5.15: Sound Library (Database button circled)

4. In the **Database** drop-down menu, select the database you created in the last exercise (e.g., my_sound_fx). You will notice that you can also select **Fairlight Sound Library** here.

5. Now that you have selected a database in **Sound Library**, you can add sounds to it. Depending upon whether you have already installed a Sound Library, you can do either of the following:

    - If you have not added any SFX libraries before, you can click the **Add Library…** button at the center of the **Sound Library** panel (*Figure 5.16*).

Figure 5.16: Add Library… button

    This **Add Library…** button will not be visible if you have already added an SFX library, in which case, use the following method to add a new one.

    - In the upper-right corner of **Sound Library** above the **Database** button, click the **Options** menu (three dots). Select **Add Library…** from the drop-down menu.

6.   In your computer's file manager, navigate to the folder that contains the SFX files that you want to import and open it.

7.   An import progress dialog box will appear as **Fairlight** indexes the audio files. Once it is finished, click the **OK** button to close the dialog box.

You have now imported your new SFX into your **Sound Library** database.

Now, we are going to search for auditions and add our newly imported SFX to the **Timeline**.

Auditioning and adding SFX to the Timeline

Now that we have imported SFX to the **Fairlight Sound Library** and created our own Sound Library, let us now add some of our imported SFX to our project:

1.   Select the audio track you want to add the SFX to by clicking on its header. The audio track header will turn from black to mid-gray to show that it is selected (*Figure 5.17*).

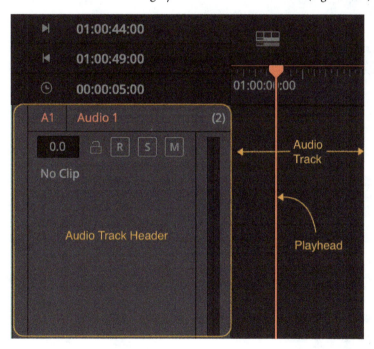

Figure 5.17: Audio Track Header

2.   Move your **Timeline** playhead (*Figure 5.17*) to where you want the SFX to start. This is usually where you see the visual source of the sound in the **Timeline** viewer, (e.g., a door closing SFX should start on the **Timeline** when a door closes in the **Viewer**).

3.   Open the **Sound Library** panel (*Figure 5.18*).

4.　In the search bar of **Sound Library** (*Figure 5.18*), type the description of the SFX you are looking for, (e.g., footsteps). Resolve will search the name and description of the SFX files and list matching SFX entries as you type your search term.

> **Top tip – searching for SFX in Sound Library**
>
> **Fairlight** uses the filename and description to match the SFX clips to your search term. This means it is important to make sure that your SFX filenames reflect their content. So, a sound recording of footsteps on gravel should be renamed from, say, Zoom000034.wav to something such as footsteps_gravel.wav.
>
> If you are unsure what to look for, you can always list all the SFX files alphabetically in the selected Sound Library database by typing three asterisks "***" into the search field in Sound Library (*Figure 5.18*). Then, you can manually scroll through them and play each one to find a sound you like.

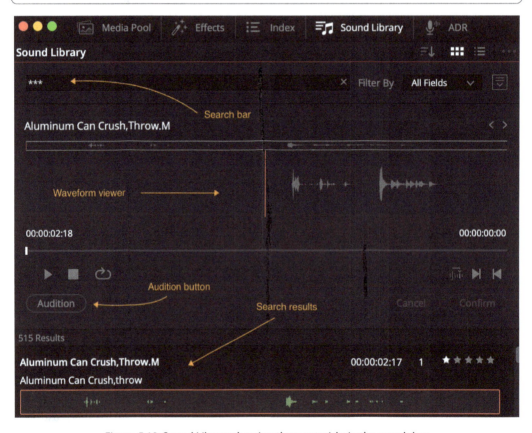

Figure 5.18: Sound Library showing three asterisks in the search bar

5. Select an SFX from the list and play it back to preview it. To play the selected SFX, do either of the following:

    · Click the **Play** button underneath the waveform viewer (*Figure 5.18*).

    · Press the spacebar on the keyboard.

6. Play each SFX in the list until you find a couple that you like. Select any one of them.

7. Click the **Audition** button (*Figure 5.18*) to temporarily place the selected SFX onto the selected track on the **Timeline**, starting at the position of your **Timeline** playhead.

8. Play the **Timeline** to hear the auditioned clip in the context of the rest of the project. To play the selected SFX on the **Timeline**, do either of the following:

    · Click the **Play** button above the **Timeline**.

    · Select the **Timeline**, then press the spacebar on the keyboard.

9. To remove the SFX clip from the **Timeline**, do either of the following:

    · Click the **Cancel** button in the **Sound Library** panel (*Figure 5.18*).

    · Select another SFX clip in the search results list.

10. Once you have found a suitable SFX clip for your project (by repeating the preceding steps), click the **Confirm** button in the **Sound Library** panel (*Figure 5.18*) to keep the selected SFX clip on the **Timeline**.

11. Click the **Sound Library** tab to hide the Sound Library (*Figure 5.18*).

You now know how to create an SFX database and use the Sound Library to import and audition your own SFX for your projects.

Plus, you have added extra audio effects plug-ins from third-party developers to expand the functionality of **Fairlight** within Resolve.

Summary

Here is what you have achieved in this chapter:

*   Adjusted the volume of different frequencies and enhanced the vocals using **Fairlight FX**

*   Imported your own AU and VST effects and applied them

*   Created a Sound Library and imported new SFX into it

*   Used the Sound Library to audition SFX and add them to the **Timeline**

You now know how to work with some audio-editing techniques on both the **Cut** and **Fairlight** pages and import audio effects and plug-ins to expand Resolve's audio-editing functionality.

Now you have completed *Part 1* of the book: *Creating Our First Project. Part 2* of the book begins with *Chapter 6*, where we will look at restoring and fixing archive footage.

In *Chapter 6*, we will also look at different file types and what video codecs are. We will look at importing and preparing different file types traditionally associated with old video footage into Resolve. We will then fix the video using FusionFX and the audio using **Fairlight**.

Questions

1. True or false? DaVinci Resolve accepts AU and VST plug-ins to add extra audio effects.

2. True or false? You are limited to only using **Fairlight FX** on the **Cut** page and need to use **Fairlight** to apply AU or VST effects.

3. True or false? The **Cut** page has access to **Fairlight FX** audio plug-ins that enable you to fix and enhance audio.

4. True or false? In order to import SFX into **Fairlight**, you need a dedicated SFX database to import the clips into.

5. True or false? An auditioned sound effect will not stay on the **Timeline** unless you click the **Confirm** button in Sound Library.

Further reading

Here is the official Steinberg website, which has some free VSTi instruments to download (scroll to the bottom of the page): `https://is.gd/5P0gli`.

Here is the link where I downloaded the free *u-he Proverb VST/AU Effects* plug-in: `https://u-he.com/products/#effects`.

Here are some websites where you can download free SFX:

- `https://freesound.org/browse/`
- `https://www.shutterstock.com/blog/household-sound-effects-room-tones`
- `https://pixabay.com/sound-effects/`
- `https://freesfx.co.uk`

Part 2: Fixing Audio and Video

In this section, we will look at fixing audio and video files in Resolve. We will start by importing archive footage and seeing how to change the properties of files to fit them into your video. We will also look at stabilizing handheld or action-cam footage. Finally, we will look at adding cut-ins or cutaways to hide jump cuts and other edits.

This section comprises the following chapters:

6

Working with Archive Footage

In this chapter, we will look at different file types and importing, as well as preparing these different file types in Resolve.

Additionally, we will look at using **Fusion FX** to restore old video footage, and change the timing of our dubbed audio to match the timing of the video. We will also use **Fairlight** to normalize the audio levels of our audio clips so that the sound volume of our video is consistent and easy for our video's viewers to listen to.

In this chapter, we're going to cover the following main topics:

- Restoring video using **Fusion FX**
- Changing audio speed
- Using **Elastic Wave**
- Normalizing audio

Let's begin our journey into working with archive footage by learning how to restore videos using **Fusion FX**.

Restoring video using Fusion FX

Often, when we use old archive videos in our projects, they can be scratchy, missing frames, and possibly have flicker. Of course, the amount of damage that will have occurred to archive footage will depend upon when it was originally recorded and how successfully it was converted into a digital format. Videos with scratches, missing frames, or flicker can be distracting to modern audiences. It can also seem out of place when mixed with modern video footage. Thankfully, Resolve has dedicated tools to restore old archive footage so that it is compatible with our audience's tastes.

To showcase just that, we will now restore a short clip from a 1920s *Buster Keaton* silent comedy called *One Week*.

In *Chapter 4*, we used **Scene Cut** to detect cuts in archive footage and added scenes to a new **Timeline**. We also dubbed in our own audio over the silent film. If you have not already completed *Chapter 4*, I highly recommend that you do so before you continue this chapter.

Open the **Timeline** named `Chapter_4_ADR` that you created in *Chapter 4,*.

Resolve FX Revival (Studio version only)

There are several useful FX plugins to restore old footage included in the Studio version of DaVinci Resolve. You can find the Resolve FX Revival plugins in the **Effects** panel on either the **Cut**, **Edit**, **Fusion**, or **Color** pages.

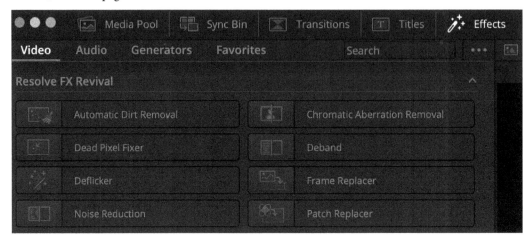

Figure 6.1: Resolve FX Revival plugins

In this example, we will use the **Cut** page to apply the Resolve FX Revival plugins to our footage.

To find the plugins, navigate to **Effects** > **Video** > **Resolve FX Revival** (*Figure 6.1*).

We will look at three of the most useful Resolve FX plugins for restoring old film footage.

These plugins are **Deflicker**, a tool to remove the flicker created by old cinema projectors; **Automatic Dirt Removal**, which can remove scratches and dirt from a frame; and **Frame Replacer**, which can replace completely damaged frames with an undamaged copy.

First, we will look at the **Deflicker** plugin, which removes the flicker seen in old film footage.

Using Deflicker (Studio version only)

When we play back Buster Keaton's *One Week* on the **Timeline**, we can see the shots have a slight flicker, as if the lights in the room are flickering. We will use the **Deflicker** plugin to remove this flicker:

1. Drag the **Deflicker** plugin onto any clip on the **Timeline**.

2. Open the **Inspector** window and select the **Effects** tab to reveal the **Deflicker** controls (*Figure 6.2*).

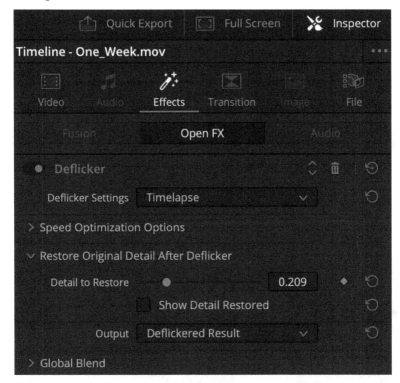

Figure 6.2: Deflicker controls in the Inspector window

3. Select **Detected Flicker** in the **Output** drop-down menu so that we can see the flicker we are removing.

4. Click on the **Restore Original Detail After Deflicker** dropdown.

5. Change **Detail to Restore** to get a balance between flicker removal and the amount of detail lost. You can do either of the following:

   - Move the **Detail to Restore** slider to the left or right

   - Type a numerical value between 0.00 and 1.00 directly into the number box, where 0.00 is no detail restored and 1.00 is the maximum amount of detail restored

6. Select **Deflickered Result** next to the **Output** settings so that we can see the result of the **Deflicker** effect.

7. Play back the clip to allow the plugin to analyze the clip and apply the **Deflicker** plugin.

8. Play back the clip again to review the results.

9. If necessary, repeat *steps 3 to 8* until the flicker has been removed.

Now that we have removed the flicker, it looks a lot better, and the film looks like it could have been shot on a modern camera.

However, there are still bits of dust, dirt, and scratches that appear on the film. Let us remove these using the **Automatic Dirt Removal** plugin.

Automatic Dirt Removal (Studio version only)

The **Automatic Dirt Removal** plugin looks for dirt in the frame by comparing the dirty frame with clean frames on either side of it. It then removes the dirt and replaces it with new pixels generated from parts of the clean image surrounding it.

Let us apply it to our clip:

1. Play back the clip after the wife calls her husband down for breakfast.
2. Note an S-shaped scratch in front of the fence to the right of the clip. We need to remove this (*Figure 6.3*).

Figure 6.3: The clip before Automatic Dirt Removal is applied

3. Drag the **Automatic Dirt Removal** FX plugin onto the clip on the **Timeline**.
4. Play back the clip to allow the plugin to analyze the clip and apply the **Automatic Dirt Removal** effect.

Play back the clip to review the results. The S-shaped scratch has mostly disappeared (*Figure 6.4*).

Figure 6.4: The clip after Automatic Dirt Removal is applied

The footage looks a lot cleaner. However, elsewhere in the video, there is still some video damage that covers a whole frame, which is where the **Frame Replacer FX** plugin can be used.

Key concept – keyframes

Keyframe is a term used in animation. The word *keyframe* comes from hand-drawn animation. Historically, a lead animator would draw the first and last frame of a movement, which were the *key* or important frames, and the junior animators would then fill in all the missing frames between those two positions.

With computer animation, the word refers to a keyframe of a video where a change is made by you (i.e., to a position of an object, e.g., point A). Further along the **Timeline,** a second keyframe is then altered by you (i.e., to a different position, e.g., point B). The computer then (like the junior animator) animates the changes between these two keyframes (i.e., the computer calculates the movement of the object between points A and B).

Resolve has a feature called auto keyframing. First, you need to add a keyframe to a frame by clicking on the gray keyframe diamond at the end of the setting. The gray keyframe diamond turns orange to show that a keyframe has been applied. This first keyframe becomes your base value for the change to start from. If you then play the video further on and change the same setting to a different value, Resolve will automatically apply a second keyframe and animate the changes between the two.

You can jump the playhead from one keyframe to another by clicking on the arrows on either side of the keyframe diamond. This means you can easily go back and tweak the settings after applying the keyframe to adjust the animation to your liking.

Most of Resolve's settings in the **Inspector** can be keyframed. Any setting that has a gray diamond at the end of the setting can be keyframed.

Frame Replacer (Studio version only)

Sometimes, a frame in the film we want to restore is so damaged that the **Automatic Dirt Removal** plugin will not be able to restore the damage. An example of this is in the video frame of the frying pan on the stove. Even though **Automatic Dirt Removal** is applied, we can still see a distortion of the stove and frying pan (*Figure 6.5*). In this situation, we can use the **Frame Replacer** plugin to delete the frame and replace it with a single frame or a blend of non-damaged frames on either side of it.

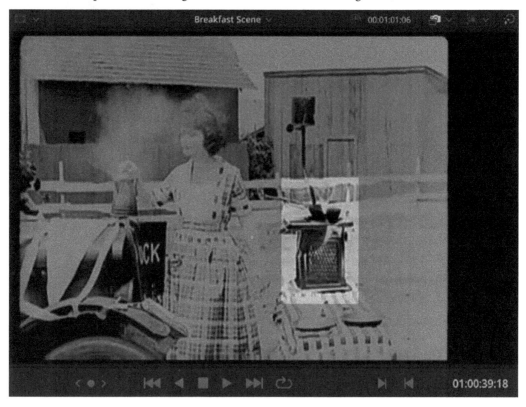

Figure 6.5: The clip with Automatic Dirt Removal applied, showing the warped stove

This is how you can do it:

1. Drag the **Frame Replacer** plugin onto the damaged clip on the **Timeline**. In my edit, it is about 01:00:03:19 on the **Timeline**.

2. Open the **Inspector** window and select the **Effects** tab.

3. Move your playhead to the frame you want to replace.

4. Check the **Replace This Frame** checkbox to select the frame you want to replace. A keyframe (*orange diamond*) is placed on this frame to mark it for removal (*Figure 6.6*).

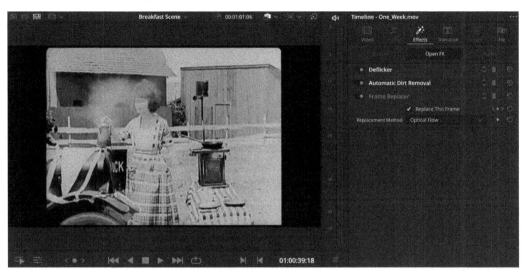

Figure 6.6: Frame Replacer applied

5. Select your **Replacement Method** option from the drop-down box to tell Resolve how to recreate the missing frame. These are the methods to choose from:

    - **From Previous**: This copies the previous frame and writes it over the current frame.

    - **From Next**: This copies the next frame and writes it over the current frame.

    - **Blend Prev/Next**: This creates a new replacement frame out of a blend of the previous and following frames.

    - **Optical Flow**: This uses DaVinci Resolve's optical flow technology to create a replacement frame from a blend of the previous and following frames. This usually creates the best results but requires more processor power from your computer. For our example with the warped stove, we will choose **Optical Flow** as our **Replacement Method**.

6. Play back the clip to allow the plugin to analyze the clip and apply the **Frame Replacer** effect.

7. Play back the clip to review the results.

As you can see from the finished result (*Figure 6.6*), **Frame Replacer** can sometimes achieve a better result than **Automatic Dirt Removal** if there is significant damage. The stove now looks brand-new!

> **Top tip – the order of FX plugins in the Inspector window**
>
> The order in which you apply any FX plugin is important, as each plugin analyzes an image based on what effects have already been applied to it. In brief, the plugins' effects are accumulative. This means you can apply several instances of the same plugin to see the effect magnified.
>
> Try applying plugins that affect the overall image first – for example, you can apply **Deflicker** and then progressively apply other plugins that remedy/enhance finer and finer detail in stages within the frame, such as **Automatic Dirt Remover**.
>
> You can always change the order of the plugins in the **Inspector** window after they have been applied by clicking on the up or down arrows at the end of the plugin name (*Figure 6.6*), to move the plugin up or down in the **Inspector** window.
>
> The plugins are applied in order, starting with the plugin listed at the top of the **Inspector** window.

Now, our footage looks a lot more visually pleasing.

As we have already covered fixing audio in *Chapter 3*, let us look at some additional audio editing techniques that will enhance our editing, such as changing the speed of the audio we dubbed in *Chapter 4* to match our actor's speech.

Changing audio speed

When we are dubbing audio, no matter how hard we try to match the original actor's lip movements, our dubbed version will still be out of sync with the original visual.

This is where it is useful to change the speed of our audio without changing the pitch. We are now going to explore how we do this on the **Cut** page.

Changing the audio speed in the Inspector

First, we'll look at changing audio speed in the **Inspector**. This is how we can do it:

1. On the **Timeline**, select an audio clip you want to change the speed for.
2. Move the clip so that the start of the audio matches the start of the actor opening their mouth in the **Viewer**.
3. Open the **Inspector** panel (*Figure 6.7*).

Figure 6.7: Speed Change controls in the Inspector

4. Click on the **Audio** tab to reveal the audio controls (*Figure 6.7*).

5. Click on **Speed Change** to reveal the speed change controls (*Figure 6.17*).

6. Keeping the **Pitch Correction** box ticked, move the **Speed %** dial (*Figure 6.7*) in either of these ways:

    *   Move it to the right to speed up the audio and hence shorten the audio clip

    *   Move it to the left to slow down the audio and hence lengthen the audio clip

7. Adjust the speed of the audio clip until the end of the audio matches the actors closing their mouth in the **Timeline Viewer**.

Changing the audio speed using the Cut Page tools

The **Cut** page has a simplified **Tools** interface that we can use to change an audio clip's speed, rather than using the **Inspector**. Changing the audio speed using the **Tools** interface has the exact same effect as using the **Inspector**; which method you choose is up to you:

1. Reveal the speed change tools by clicking the **Tools** icon (white slider bars) in the toolbar under the **Viewer** above the **Timeline**.

Figure 6.8: Cut Page tools with Speed Change selected

2. Click on and drag the **Speed** slider (highlighted by an orange box) left to reduce the audio speed or right to increase it. A value of 1.00 is the original speed, 1.50 is one and a half times the original speed, and 0.50 is half the original speed.

    Once any change has been made, an orange dot will appear above the icon of the tool that has been used, indicating that a change has been applied to the original clip (*Figure 6.8*).

3. Click on the **Tools** icon again to hide the tools.

Now, the audio better matches the speed of the actor's speech. However, there will be words in the middle of their speech that do not match. So, let's fix this by using the **Elastic Wave** tool in **Fairlight.**

Using Elastic Wave

Elastic Wave is a keyframe-based method of changing the timing of audio without changing the pitch. Audio keyframes are placed at points in the audio clip where you do not want the audio waveform to move or stretch. The audio between the keyframes can be stretched or compressed to then fit key timings in the actor's performance:

1. To enable **Elastic Wave**, you first need to go to the **Fairlight** page.

2. Right-click on the audio clip you want to enable **Elastic Wave** for and choose **Elastic Wave** from the pop-up menu. This will reveal the **Elastic Wave** controls, shown by the words **Elastic Wave** now written across the top of the audio clip (*Figure 6.9*).

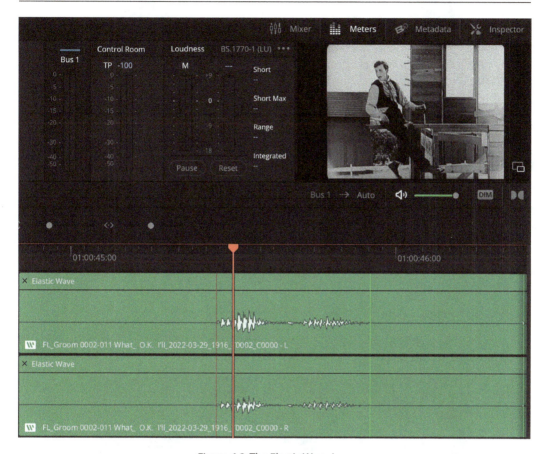

Figure 6.9: The Elastic Wave icon

3. *Press Command (CMD)* + click on the audio waveform where you want to add a keyframe. A green vertical line will appear on the waveform to show where the audio keyframe has been applied.

    You will also see a **W** icon at the bottom left of the audio clip to show that **Elastic Wave** has been enabled (*Figure 6.9*).

4. Apply audio keyframes at the beginning and end of the audio waveform you want to shorten or lengthen the timing of (*Figure 6.9*).

5. Drag the keyframe to either end of the waveform to lengthen or shorten the audio waveform to match the length of time the actor is speaking in the **Viewer**. The keyframe will turn red to show which one you have selected for adjustment (*Figure 6.9*).

6. Close the **Elastic Wave** controls by double-clicking on the **W** icon at the bottom left of the audio clip.

If you have already made a speed change to the audio clip elsewhere, such as on the **Cut** or **Edit** pages, you will notice that when you go to **Fairlight**, the **Elastic Wave W** icon has already been applied to the audio clip. In this case, you do not need to follow *step 2* in the previous steps; instead, double-click on the **Elastic Wave W** icon to reveal the **Elastic Wave** controls, and then follow *steps 3 to 6* as usual.

We have now adjusted our dubbed audio to match the lips of the actor in the video clip talking. We just need to make sure that all our audio clips are at a consistent volume level.

Normalizing audio

Staying on the **Fairlight** page, we will now use Resolve to automatically change all the audio levels of the clips to a specified maximum volume. Normalization stops drastic changes in the peak volume, which can deafen your viewer and have them constantly reaching for the volume control. You can also access the **Audio Normalization** controls in the same way on the **Edit** page.

Rather than individually adjusting the volume of each clip to match each other, you can instruct Resolve to automatically adjust the clips so that the peak (loudest) volume of each clip is the same.

This automatic volume leveling of the clips is called *normalizing audio*:

1. To do this, right-click the selected audio clip or clips you want to normalize and select **Normalize Audio Levels...** from the pop-up menu.

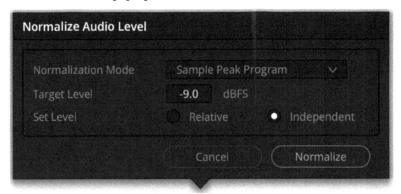

Figure 6.10: The Normalize Audio Level control pop-up box

2. This reveals a **Normalize Audio Level** control pop-up box (*Figure 6.10*) with the following options:

    *   **Normalization Mode**: This is the method used by **Fairlight** to analyze audio. The default **Sample Peak Program** option is fine for most uses, particularly social media. The other normalization modes are more useful when delivering audio for television broadcasts or streaming services, such as Netflix.

    *   **Target Level**: This is the target volume level that the normalization will match. This is measured in **decibels relative to full scale (dBFS)**.

3. **Set Level**: This option only appears when you have selected two or more clips on which you will normalize the audio. Selecting **Relative** will change the audio across all the clips collectively to be relative to the clip with the highest audio peak. In practice, all the clips together are treated as one clip. So, for example, if the loudest clip needs to be reduced by 10 dBs, then all the clips will be reduced by 10 dBs. Selecting **Independent** will change the audio of each clip independently to meet the **Target Level** setting. Therefore, some clips will need to be lowered and others raised to meet the dBFS **Target Level** setting. This is useful if you want to ensure all the clips have similar volume levels.

Key concept – dBFS

dBFS is where 0 is the maximum volume of your audio system before distorting or clipping occurs. A value of -12 dBFS is 12 dB lower than the system maximum. Hence, the larger the negative number, the quieter the sound will be.

For example, a dBFS value of -12 (e.g., human speech) will be quieter than a dBFS value of -9 (e.g., shouting).

Common target level values:

- Human speech: -12 dBFS

- SFX: -10 dBFS to -30 dBFS

- Music: -20 dBFS to -30 dBFS

---

4. Change your **Target Level** to -12 dBFS as a rough target for human speech.

5. Click the **Normalize** button.

6. Note that the waveforms of the selected audio clip(s) will now become higher or lower to match the selected dBFS Target Level.

We have now normalized all our audio clips to reach a target level of -12 dBFS so that it is easier for our audience to hear what is being said without being deafened, or having to strain to hear what is being said.

We have now restored our archive video footage, matched our dubbed audio to the actor's performance, and leveled out any variance in the sound levels of our recorded audio clips.

With this, we draw *Chapter 6* to a close.

Summary

Here is what you have achieved in *Chapter 6*:

- Used a range of **Fusion FX** tools to restore old footage

- Changed the speed of audio clips

- Used **Elastic Wave** to change the timing of dubbed audio clips to fit your video

- Normalized audio levels in **Fairlight**

In *Chapter 7*, we will look at different tools we can use to stabilize video footage in the **Cut**, **Edit**, and **Color** pages. We will use these tools to stabile our Buster Keaton archive footage from *One Week*.

Questions

1. True or false? You need the Studio version of Resolve to use **Fusion FX**.

2. True or false? You can only change the speed of audio clips on the **Fairlight** page.

3. True or false? **Elastic Wave** controls can only be accessed on the **Fairlight** page.

4. True or false? Normalizing audio can only be done on the **Fairlight** page.

Further reading

Archive footage websites:

- `https://archive.org/details/movies`

- `https://media.nationalarchives.gov.uk/index.php/category/video/`

- `https://www.britishpathe.com`

- `https://www.ap.org/content/video/archive-video`

- `https://publicdomainreview.org/collections/film`

7
Stabilizing Shaky Footage

In *Chapter 6*, we looked at common methods of restoring archive footage using Resolve FX. A typical problem that can plague both archive and modern footage is shaky video. In this chapter, you will learn how to stabilize shaky footage in the **Cut**, **Edit**, and **Color** pages, which you can apply to both your archive and recent footage.

In this chapter, we're going to cover the following main topics:

- How software-based video stabilization works

- Stabilizing video on the **Cut** page using basic controls

- Stabilizing video with more advanced controls in the **Inspector** window on the **Cut** and **Edit** pages

- Stabilizing video on the **Color** page using the **Tracker** palette and the **Classic** tracker

Although it is always preferable to capture stable video footage with our camera, this is not always practical. For example, we may be creating a *run and gun*-style documentary where there is no time or space to attach expensive stabilization equipment to our camera. It could also be that the batteries on your hand-held camera stabilizer failed, or you filmed the action on a mobile phone with no tripod for support.

Whatever the reason for having shot shaky video footage, there are times we need to fix the shaky footage post-production to make it easier for the viewer to watch. DaVinci Resolve has an excellent stabilizer that you can use to convert unstable footage into footage that could have been created on a tripod.

Technical requirements

For the exercises in this chapter, we will be starting a new project where we stabilize a wedding video.

You can download the `The Wedding.DRA` file here, which contains the wedding video media required for the next few chapters (The *Chapter 7* folder): `https://packt.link/B5bqz`.

How video stabilization works

It is important to understand the basic principles of how video stabilization works in post-production software so that we can understand how to use the controls to fine-tune the software's video stabilization.

The steps of video stabilization

As with all editing software stabilizers, Resolve will compensate for the camera movement by cropping into the video. So, as with all video stabilization, there will be a slight loss of resolution and the image will be slightly zoomed-in, due to the need to crop the image. Let us look at how and why video stabilization crops the image.

1. The first step in video stabilization is to enable the software to detect and track movement in our video clip. We can select which type of movement the software detects by selecting options such as **Perspective**, **Similarity**, and **Translation**. These movement detection options will be described in further detail in the following sections.

2. The next step is for the software to stabilize the clip by moving it in the opposite direction to the camera movement to counteract the shaky movement.

    So, for example, if the camera moves to the left, causing the subject to be placed to the right of the frame, the software will move the clip to the left to keep the subject of the video in its original place (*Figure 7.1*).

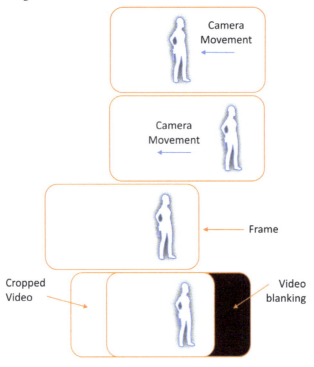

Figure 7.1: Software compensating for camera movement

Unfortunately, this leaves a black edge (video blanking) within the frame, as there is no video information to show because we have moved the clip within the frame.

3. So, the final step is for the software to crop the clip by zooming in to remove the black edges.

This is difficult to illustrate using drawings, so let's demonstrate this using Resolve itself. You can use your own shaky footage for this exercise or use the wedding video footage in the The Wedding. DRA file, which we will continue to improve in *Chapter 8*

Creating a new Timeline and reviewing our footage

This is how you can create a new **Timeline** and review the footage:

1. Let us set up our new project and review the footage before we stabilize the video. Import The Wedding.DRA into Resolve, as we covered in previous chapters.

    The footage is already separated into **bins** for the **Timelines** and video files. You can use the **Timelines** to compare your version with a final version.

2. Create a new **Timeline** in the **Timelines** bin by right-clicking in an empty section of the bin and selecting **Timelines > Create New Timeline...**; rename it to something appropriate (e.g., **My Wedding**) following the steps to create a **Timeline** we outlined in *Chapter 4*.

3. Before we begin our edit, let us watch through the footage in the **Media Pool** to become familiar with it using the media playback techniques we have covered previously.

Now that you have quickly reviewed your footage, note that there are some shaky camera moves as well as out-of-focus shots that we need to remove. We will need to replace it with stable footage.

Stabilizing our video footage

As it is wedding video footage, we cannot reshoot the video without asking the bride and groom to repeat the whole ceremony, so we will need to be able to fix what we have. Thankfully, Resolve has some easy-to-use tools that help us stabilize existing footage:

1. Drag the clip named A001_04041646_C033.mov from the **The Wedding.dra > MediaFiles** bin onto the **Timeline** on the **Edit** page.

2. Select the clip on the **Timeline** you want to stabilize (e.g., A001_04041646_C033.mov) and open the **Inspector** window.

Figure 7.2: The pre-stabilized clip

3.   Under the **Video** tab, double-click **Stabilization** to reveal the stabilization controls.

4.   Select **Translation** in the **Mode** drop-down menu. This will limit Resolve to only looking for up and down (tilt) and left and right (pan) camera movements and ignore any other movement, such as rotation or zooming. This is so that we can see how Resolve stabilizes a shot on the 2D plane without being distracted by the depth of movement.

5.   Leave the **Cropping Ratio** slider in the middle to show a default value of 0.500. This will enable the cropping of the clip so that we can see the black edges, as a result of Resolve moving the clip. A value of 1.000 will mean that no cropping will take place, which results in no stabilization. A value of 0.250 will enable maximum cropping.

6.   Leave the **Smooth** slider all the way to the left to show a default value of 0.250.

7.   Leave the **Strength** slider at full strength, at its default value of 1.000.

8.   Untick the **Zoom** checkbox to reveal the extent of the crop that Resolve will apply.

9.   Click the **Stabilize** button to stabilize the clip.

Now that we have stabilized the clip with **Cropping Ratio** reduced, note the black edges added where the video used to be. In this example, there are black edges to the left and bottom of the frame, which have cropped the top and right of the original video's frame. So, in this instance, Resolve has moved this frame up and to the right to counteract the camera shake.

Figure 7.3: The stabilized clip with black edges (the left and bottom edges)

Now, to demonstrate how the software zooms in to crop out the black edges, select the **Zoom** checkbox.

Figure 7.4: The zoomed-in stabilized clip with no black edges

Note that although the clip is now stabilized, the clip has also zoomed in. This will result in the subject taking up more of the frame, some of the background being cropped out, and a slight loss of video resolution.

For this reason, it is better to use software stabilization sparingly and record video as stably as you can.

Otherwise, you can record video at a higher resolution than your video's intended final output resolution and frame the video so that there is room to crop the clip later. For example, shooting in 4K to deliver a 1080p video will allow room for cropping without a loss in video resolution.

Now that we know how software stabilization works, let us look at the different ways that you can stabilize video using DaVinci Resolve, starting with the **Cut** page. You can always try these different techniques on the wedding video footage to see how each one works.

Stabilizing video on the Cut page

As discussed at the start of the chapter, there are many reasons for needing to fix shaky footage in post-production.

Whatever the reason for the shaky footage, DaVinci Resolve has an excellent stabilizer you can use to convert unstable footage into footage that could have been created on a tripod.

As with all editing software stabilizers, Resolve will compensate for the camera movement by cropping into the video. So, with all video stabilization, there will be a slight loss of resolution and the image will be slightly zoomed-in, due to the need to crop the image.

Resolve has many different ways to stabilize video footage.

First, let us look at how we can stabilize video footage on the **Cut** page:

1. Select the clip on the **Timeline** that you want to stabilize.

2. Click on the **Tools** button (the sliders in *Figure 7.5*) to reveal the **Clip** tools.

Figure 7.5: The Tools button

3. Click on the **Stabilize** icon (which resembles a *shaky camera*) to reveal the stabilization controls (*Figure 7.6*).

Figure 7.6: The Stabilize icon

4. Choose your stabilization method in the drop-down menu box (*Figure 7.6*). The options are listed from the strongest stabilization method (**Perspective**) to the least complex method (**Translation**):

    *   **Perspective**: This option enables Resolve to analyze the footage for changes in perspective, pan, tilt, zoom, and rotation and compensates for this movement.

    *   **Similarity**: This option is the same as **Perspective** but without analyzing for changes in perspective. Therefore, this stabilization method only compensates for changes in pan, tilt, zoom, and rotation. This is a good option to choose if the **Perspective** option caused unwanted motion artifacts.

    *   **Translation**: This option only analyzes the video for changes in pan and tilt movements. It ignores changes in perspective, zoom, and rotation. If the **Similarity** option did not work, then choosing **Translation** instead may be a better option, as it limits the stabilization to just up-and-down and left-and-right movement on a 2D plane parallel with the camera, ignoring any movement in depth such as zooming. Of course, choosing **Translation** is a good option if you want to stabilize a shot without affecting the intentional zoom of the camera lens.

5. Click on the **Stabilize** button to stabilize the clip using the method you have just selected. Resolve will analyze the clip for movement (*Figure 7.7*) and apply stabilization to remove the movement, based on the stabilization method you chose.

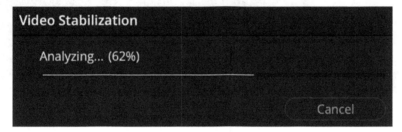

Figure 7.7: The Video Stabilization progress bar

If you do not like the results of the stabilization method you selected, you can select a different stabilization method and click again on the **Stabilize** button to get Resolve to re-analyze and stabilize the clip, based on the newly selected method.

The stabilization tool used on the **Cut** page is exactly the same tool that was used on the **Edit** and **Color** pages. The main difference is that there are more controls available on the **Edit** and **Color** pages to fine-tune the stabilization result.

One place to reveal more stabilization controls is in the Inspector on both the **Edit** and **Cut** pages.

Stabilizing video in the Inspector on the Cut and Edit pages

Let us now look at the stabilization controls in the video stabilizer, which can be found in the **Inspector** panel on both the **Cut** and **Edit** pages. As the stabilization controls are exactly the same, we will stay on the **Cut** page to look at these extra controls in the Inspector:

1. Select the clip on the **Timeline** that you want to stabilize.

2. Click on the **Inspector** tab to reveal the **Video Inspector** controls.

3. Double-click on the **Stabilization** tab to reveal the **Stabilization** controls (*Figure 7.8*).

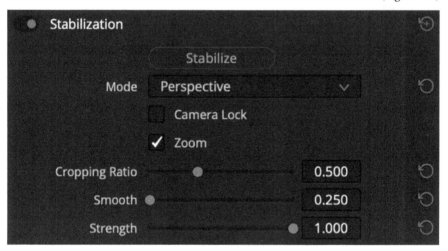

Figure 7.8: The Stabilization section of the Video Inspector panel

4. The **Mode** drop-down menu allows you to select your stabilization method, just like we did with the other simplified stabilization controls on the **Cut** page.

5. The extra controls in the **Inspector** window, which are not in the basic controls on the **Cut** page, allow us to control the power or amount of stabilization. Every time you change one of these settings, you will need to click the **Stabilize** button again to apply the change:

    *   **Camera Lock**: Selecting this checkbox disables the **Cropping Ratio**, **Zoom**, and **Smooth** controls. The stabilizer will attempt to lock the shot as if the video was filmed on a *locked-off* tripod where there is absolutely no camera movement.

- **Zoom**: Selecting this checkbox zooms the video in to eliminate the black edges created by the stabilization:

  - The **Cropping Ratio** slider (with values from 0.250 to 1.000) determines how aggressive the stabilization will be and, hence, the amount of zooming that Resolve will need to apply. A value of 1.000 means that no zooming will occur. The lower the number, the more zooming will occur.

  - **Smooth**: The **Smooth** slider (values 0.250 to 1.000) determines how much smoothing to apply to the original camera movement. A value of 1.000 means that maximum smoothing will be applied and very little camera movement will be retained, and a low value will allow more of the original camera movement to be applied to the clip.

  - **Strength**: This slider (values -1.000 to 1.000) allows you to choose the strength of the stabilization, where 1.000 is maximum stabilization and 0.000 is no stabilization. A value under 1.000 and above 0.000 allows some of the original camera movement to be mixed in with the stabilization. A negative value is used in more advanced match moving of shots (where two video clips are overlaid and need to have the same video movement), so this is safe to ignore for the majority of users.

  The **Smooth** and **Strength** controls are used together to get a balance of smooth stabilized shoots while retaining a blend of some of the original camera movement. Try different combinations of the **Smooth** and **Strength** settings until you get the result you like.

As you can see, the two different video stabilization controls on the **Cut** page do not differ in function; the Inspector on the **Edit** and **Cut** pages just has more advanced controls to tweak the strength of the stabilization applied to the video clip.

Let us now look at how video stabilization works differently on the **Color** page.

Stabilizing video on the Color page

Stabilizing video on the **Color** page uses the same tools as the **Cut** and **Edit** pages, but the interface is very different. As the **Color** page interface is so different and can be confusing for even advanced editors, we will only look at how the **Color** page stabilizes video and not the vast amount of other tools we can access there:

1. Place the playhead over the clip you want to stabilize on the **Timeline**.

2. Go to the **Color** page; the clip you selected on the **Timeline** will already be selected.

3. Open the **Tracker** palette (*Figure 7.9*).

Figure 7.9: The Tracker palette

4. Select the **Stabilizer** icon (which looks like a camera shaking in *Figure 7.9*).

5. The drop-down menu at the bottom right of the palette (*Figure 7.9*) allows you to select your stabilization method, just like we did on the **Edit** and **Cut** pages.

6. Along the bottom of the **Tracker** palette (*Figure 7.9*) are the same stabilizer controls from the **Inspector** window on the **Edit** page. They work in exactly the same way. The only difference is that the interface for the controls looks and operates slightly differently.

7. Click the **Stabilize** button (*Figure 7.9*).

Resolve will now analyze the clip with camera movement and track the movement.

The clip will be stabilized, just like on the **Cut** and **Edit** pages. However, uniquely to the **Color** page, the tracking data will be represented in a graph (*Figure 7.9*) by four colored (pink, green, blue, and yellow) lines:

-   The *pink* line shows the **Zoom** tracking data

-   The *green* line shows the **Pan** tracking data

-   The *blue* line shows the **Tilt** tracking data

-   The *yellow* line shows the **Rotate** tracking data

If the line is completely flat, this means that there was no tracking data recorded.

Numerical tracking data is also shown for each color graph by the corresponding color number at the bottom right of the graph (*Figure 7.9*).

> **Top tip – more than one way to change a number**
>
> Most number entry boxes in Resolve act like sliders when you change the numerical value in them.
>
> If you hold your left mouse button down and drag left or right, you will change the number values in the box up or down, just like you would with a visible slider interface.

We know that selecting **Translation** only tracks the **Pan** and **Tilt** data to stabilize the shot. Let us see what the graph shows when we stabilize a shot using the **Translation** mode. In theory, only the green and blue lines should show any movement, and the pink and yellow lines will remain completely flat:

1. On the **Color** page, open the **Tracker** palette.

2. Click the **Stabilizer** icon.

3. Select the **Translation** mode.

4. Click the **Stabilize** button.

You should now see that the pink and yellow lines are completely flat, and there will be some movement up and down the blue and green lines.

To see how the different stabilization modes work, select them on the **Color** page and compare the results of the graphs.

> **Top tip**
>
> Any tracking that is performed on the **Cut** and **Edit** pages will show on the **Color** page.
>
> So, if you performed a track on the **Edit** page, you could still go to the **Color** page and see the results of the tracking data in the **Tracker - Stabilizer** graph without having to re-stabilize the clip on the **Color** page.

Now, if you really want fine control of stabilization in Resolve, you can select **Classic Stabilizer**. The **Classic Stabilizer** was the main way to stabilize video footage in DaVinci Resolve version 12.5 and earlier. With DaVinci Resolve version 14 onward, you have the simplified stabilizers, as described earlier in this chapter, where you only have three stabilization presets to choose from, **Perspective**, **Similarity**, or **Translation**. The **Classic Stabilizer** is still included, as it has the option to individually select each of the **Pan**, **Tilt**, **Zoom**, and **Rotate** options and turn them on or off, allowing you to have more control over your stabilization. If using the new stabilization controls doesn't work, you can always try your own combination of stabilization using the **Classic Stabilizer**. Let us look at how the **Classic Stabilizer** works.

The Classic Stabilizer

The **Classic Stabilizer** separates stabilization into three distinct steps. So far, all the techniques we have covered on the **Cut**, **Edit**, and **Color** pages combine these three steps.

Overall, these steps are as follows:

1. The first step is to analyze the clip for movement.
2. The second step is to select what stabilization settings to apply.
3. The third step is to stabilize the clip.

Let us see how this works in practice using the default **Cloud Tracker** settings.

Classic stabilizing using the Cloud Tracker

The Classic Stabilizer by default generates a cloud of tracking points to help us stabilize our video footage:

1. In the **Tracker - Stabilizer** palette, select the **options** button (looks like a row of three dots ...).
2. Select **Classic Stabilizer** in the drop-down menu (*Figure 7.10*). A tick will show next to it in the menu to show that it is selected.

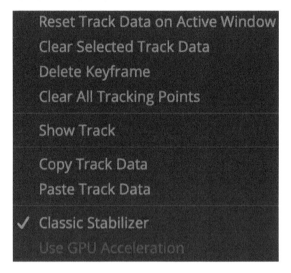

Figure 7.10: The Tracker options' drop-down menu

The **Tracker** palette interface will change to show different tracking controls for **Classic Stabilizer** (*Figure 7.11*).

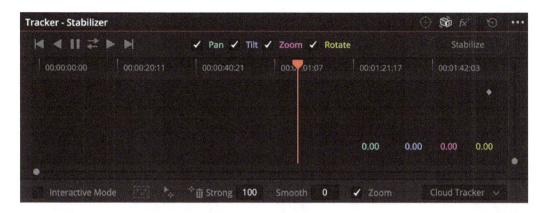

Figure 7.11: Classic Stabilizer

3. Analyze the clip for movement as follows:

    I.   Select the movement you want Resolve to look for by selecting any combination of the **Pan**, **Tilt**, **Zoom**, and **Rotate** checkboxes, located at the top of the **Classic Stabilizer** palette.

    II.   Start tracking the clip for movement by selecting one of the following:

    •   The **Track Forward** button (looks like a play button)

    •   The **Track Forwards and Reverse** button (looks like arrows pointing in both directions)

    •   The **Track Reverse** button (looks like a reverse play button)

4. Select the stabilization settings. Like in the previous stabilization methods, you can choose the **Strong**, **Smooth**, and **Zoom** settings. There are no options for **Camera Lock** or **Cropping Ratio**.

5. Click the **Stabilize** button to perform the stabilization.

When analyzing the clip for movement, by default, Resolve uses a **Cloud Tracker**, which creates a cloud of tiny white crosses (you can see them on the bride's dress in *Figure 7.12*) on distinguishing features in the starting frame of the video. These crosses are then tracked for changes in their position within the frame.

Figure 7.12: Tracking points

You can select a **Point Tracker** instead of the **Cloud Tracker** if you want more control over what points in the video you want Resolve to track.

This is how you use a Point Tracker.

Classic stabilizing using the Point Tracker

To make Resolve track and stabilize using a selected point in the video, we need to follow these steps:

1. Select **Point Tracker** in the **Cloud Tracker** drop-down menu.

2. Click on the **Add Tracker Point** icon to add a tracker point (*a blue cross*) to the current frame shown in the **Viewer**.

3. Click and drag the **Tracker Point** to reposition it to the point in the frame you want to track. When you click on the tracker point, it will turn *red* to show that you have selected it.

4. Repeat *steps 2* to *3* until you have positioned as many tracking points as you need.

    Now, complete the remaining *steps 5* to *7* as if you were completing a cloud stabilization (analyze the clip, select the stabilization settings, and stabilize).

5. Analyze the clip for movement exactly as we did in *step 3* for the Cloud Tracker.

6. Select the stabilization settings exactly as we did in *step 4* for the Cloud Tracker.

7. Click the **Stabilize** button to perform the stabilization.

Top tips – the Point Tracker

When selecting a point to track, select a high-contrast point that is easily identifiable from the rest of the clip that represents the motion that you want to track. Select at least four points if you want to stabilize **Pan**, **Tilt**, **Rotation**, and **Zoom**.

Keep the tracker points at the same depth from the camera as the subject being tracked, as points in the background move slower than points in the foreground. Also, choosing several points on the object being tracked will help the track be more successful. Do not put tracker points on different objects moving in different directions, as this will confuse the Point Tracker.

If you apply too many tracking points, you can delete some by selecting one (turns *red*) and then clicking on the **Delete Tracker Point** icon (the *trash can*).

When adding tracker points, they will all appear in the center of the frame. This means that if you click the **Add Tracker Point** icon several times, they will all stack upon each other, making it appear as if there is only one tracker point. Move the topmost tracker point, and you will see more underneath.

Now, we know how to stabilize a video in Resolve using stabilization controls on the **Cut**, **Edit**, and **Color** pages. We also understand how stabilization works and how to add tracker points manually. This brings us to the end of the chapter.

Summary

You now know the range of stabilization tools available in DaVinci Resolve, whether it's a quick and easy video stabilization on the **Cut** page or a more involved stabilization on the **Color** page. You can now choose the best stabilization for the needs of your footage, always knowing that you have access to some advanced stabilization tools should you ever need them.

Here is what you have achieved in *Chapter 7*:

- You learned how video stabilization works in editing software
- You learned how to stabilize footage on the **Cut** page
- You learned how to stabilize footage on the **Edit** page
- You learned how to stabilize footage on the **Color** page using both **Cloud** and **Point** trackers

In *Chapter 8*, we will look at using cutaways and cut-ins on the **Cut** page to hide bad edits. We will also look at how to do this on the **Edit** page.

Questions

1. True or false? When shooting footage, it is always better to stabilize footage in post-production rather than worry about capturing stable footage.

2. True or false? There is just one way to stabilize a clip on the **Cut** page.

3. True or false? Stabilization on the **Edit** page works completely differently from the **Cut** page.

4. True or false? There is more than one way to stabilize a clip on the **Color** page.

Further reading

An overview of the different types of stabilization for video production and post-production: `https://moviola.com/technique/dslr-filmmaking-stabilizing-techniques-production-and-post/`

8
Hiding the Cut – Making Our Edits Invisible

In this chapter, we will look at how to use cutaways and cut-ins on the **Cut** page to hide bad edits. Additionally, we will look at how to hide jarring cuts on the **Edit** page using a split edit. Finally, we will look at how to use the **Smooth Cut Transition** to make a cut invisible.

In this chapter, we are going to cover the following main topics:

- What cutaways and cut-ins are and how to use them
- Continuity editing
- Creating cut-ins
- Creating cutaways
- Using Smooth Cut to hide an edit
- Hiding cuts with a split edit

So far, we have only really covered the basics of editing, such as introducing basic cuts to trim our content down. However, editing is more than just cutting out unwanted video footage. It is also a way to help tell a story seamlessly.

Our task as an editor is to choose clips and put them together in a sequence that helps tell a story. To ensure our audience can connect with the story we are telling, it is important that the editing process itself is not noticeable, as this can distract from the content of the video.

One editing technique that both helps tell the story and hides obvious or jarring edits is the use of cut-ins and cutaways inserted from B-roll video footage.

Technical requirements

To follow along with the exercises in this chapter, download the wedding footage (The_Wedding. DRA), as described in *Chapter 7*, and stabilize the clip named A001_04041646_C033.mov if you haven't done so already. You can download the required video clips from: https://packt. link/B5bqz Before we begin our first exercise, we need to explain some editing terms such as cutaways, cut-ins, and B-roll.

Understanding cutaways, cut-ins, and the editing process

To understand what B-roll is, let us uncover where the term first came from.

In the film industry, the main camera team (the main unit) shoots the central action of the film's key characters with *Camera A*, and a second camera team (called the second unit) will shoot other footage with *Camera B*, where you do not need the main actors present. The footage from Camera A is called A-roll, and the footage from Camera B is called B-roll. Both were named in the early days when movies were exclusively shot using rolls of film (*Figure 8.1*):

Figure 8.1: Image of rolls of film (LDGE at English Wikipedia): https://en.wikipedia.org/wiki/

Since B-roll from the second unit is used to capture elements of the story where the main actor is not needed, this frees up time for the main unit to focus on the main story, while the second unit captures other shots for the edit such as cut-ins or cutaways. Cut-ins and cutaways are two types of B-roll footage that can be used by the editor to hide an obvious edit, which we will demonstrate in a later exercise where we will edit a short wedding video.

So, what is the difference between a cut-in and a cutaway? Let's find out together.

Cut-ins

Generally, cut-ins are close-up shots of the main action within the scene.

For example, if the main actor is making an origami bird, then the second unit would film a tight close-up shot of a hand double who is an origami expert (with similar hands to the lead actor) making the origami bird. This would be a cut-in as it illustrates the main action of the central character; hence, the name *cut-in* as we are "cutting in" to the main action.

Cutaways

Cutaways are different from cut-ins in that they *cut away* from the main action of the key actor in the scene. Cutaways can often show other visual information that can help illustrate the main action of the scene or give a different interpretation to it.

For example, the main actor might be making an origami bird; however, if we cut away to a ticking clock, then we will assume that there might be a time limit to the origami, such as a contest. If using the same example, where we cut away to a Japanese serenity garden, we would assume that the actor is making the origami bird to unwind or relax.

Now that we understand the difference between cut-ins and cutaways, let us look at the overall editing process and how cuts are not just used to hide mistakes in the footage.

The editing process

A good cut-in or cutaway will not only be used to hide an unwanted edit, but also to help enhance or tell the story better.

Some examples of how they can be used to enhance a story are listed as follows:

- *Extend or slow down time*: Cutting out surplus footage such as long pauses, and covering the edits with cutaways, cut-ins, or shots from a different camera angle, can speed up time. On the other hand, adding cuts into the footage, without removing any footage, and adding cutaways, cut-ins, or other angles will slow down time.

- *Show the internal thoughts of the main character in the scene*: A cutaway can be used to illustrate what the character is thinking about, such as a flashback to a childhood memory.

- *Create suspense or tension*: Cutting away to an image of a threat (such as an approaching train or tidal wave) that the main character is oblivious to can create tension in the audience. Alternatively, cutting into the main character holding a weapon under the table can add tension to the conversation being had above the table.

- *Reveal a comedy gag or punchline*: The main character might be saying something profound but we cut away to something to illustrate that what they are saying lacks any credibility, causing the audience to laugh. One example would be the main character saying that they were cool and collected in an emergency and we then cut away to them panicking in the emergency.

- *Add visual interest*: If we filmed the whole scene with one shot, often, the viewer can get bored and lose interest. Adding a shot of the same scene but from another angle can keep the viewers' attention by adding visual interest.

Now we are going to illustrate how to use cut-ins and cutaways to both enhance the story in a wedding video as well as hide some poor footage.

Understanding continuity editing

One use of cut-ins and cutaways is to help with **continuity editing**. Continuity editing is an editing technique where we aim to maintain the timing of the clips on the **Timeline** as if the action is happening in a continuous moment in time (even if the clips were shot at different times or on different days). Rather than try and understand it with a definition, it is easier to demonstrate it in the following exercise using the *The_Wedding* project that we downloaded earlier.

Even though we have stabilized the wedding video footage, there are still parts of the clips that we cannot fix, such as the camera drifting in and out of focus.

We will need to cut these out and hide the resulting **jump cut** with other B-roll footage, which was shot at a different time by a wedding guest. I have already color-coded the B-roll footage we are going to use as a teal color and the master footage as a lime green color to make them easier to find and differentiate. If you want to change the clip colors yourself, you can do so by right-clicking on any clip, and from the drop-down menu, selecting **Clip Color**, and choosing your color.

Key concept – jump cut

A jump cut is where a piece in the middle of the same shot is cut out, leaving a jump in time.

For example, say a person walks from left to right in a single shot, cutting out the middle part of the shot between the start and the end of the walk will create a jump in time. As we don't see them move between the start and end point of their walk, this will create a shortening of time for the scene.

Jump cuts are often seen in YouTube tutorial videos where the subject has edited out their mistakes (for example, "umms" and "ahhs") but keeps talking in the same shot.

Jump cuts are often avoided as they throw the viewer off watching the video due to the jarring effect they create. However, sometimes, they can be used for good effect if you want the viewer to feel uneasy, such as in horror or action sequences during a film.

Now that we know more about jump cuts and clip colors, we will jump into the exercise. Using our stabilized wedding video clip in the **Timeline** of the **Edit** page, we are going to remove the shaky and out-of-focus camera moves and replace them with some B-roll footage and make it seem that it was all shot in the same moment:

1. Move the playhead on the **Timeline** to timecode 01:00:37:02.

2. Add a cut using the **Razor** tool (the *CMD* + *B* key combination is the shortcut).

    A cut will appear under the playhead on the **Timeline**.

3. Move the **Timeline** playhead to the following timecode: 01:00:45:00.

4. Add another cut (*CMD* + *B*).

5. Select this new clip in-between the two cuts (between 01:00:37:02 and 01:00:45:00) and delete it by pressing the *Backspace* key on your keyboard.

    This has created a gap in your **Timeline** on the Video 1 Track, which we need to cover (*Figure 8.2*):

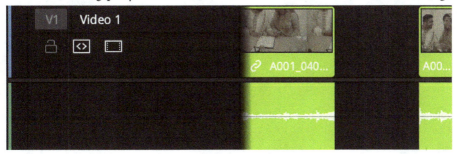

Figure 8.2: Gap in the Timeline

Top tip – deleting clips

The two easiest ways to delete a clip on the **Timeline** are to select it and then use either the *Delete* or *Backspace* key on your keyboard to remove it. However, these two keys delete the clip in different ways and have different effects on your edit.

The *Backspace* key will delete a clip and leave a gap in the **Timeline**. This is good if you want to leave the timing of the video intact without creating a jump cut. The clip can be easily covered over with another clip from the B-roll, as we demonstrate in this chapter.

The *Delete* key will delete the clip and ripple all the other clips along the **Timeline** to close the gap; this is called a **Ripple Delete**. This will create a jump cut if it is used to delete the middle section of a clip. A **Ripple Delete** is best used when removing clips that are no longer needed to shorten the length of the video.

---

6. Go to the B-roll bin in the **Media Pool** and find a close-up clip of the groom's hands signing the wedding certificate (i.e., A001_04041647_C034.mov). Double-click on it to see it in the **Viewer**.

7.   This clip is too long to use in its entirety, so let us shorten it before we add it to the **Timeline**. We also need to find a part of the clip that matches the action of the existing clips on the **Timeline**.

The end of the first clip on the **Timeline** has both the bride and groom looking down at the paper they are about to sign. We need the start of the new clip in the **Media Pool** to start with the couple in a similar position.

Add an In point (where the clip starts) by pressing the *I* key on your keyboard at 16:51:05:15 (*Figure 8.3*). A gray dot will appear under your Viewer's playhead to mark the **In point**:

Figure 8.3: In and Out points marked up on the source clip in the Viewer

Key concept – In and Out points

We used In and Out points when using **Fairlight** to dub audio in *Chapter 4*. In and Out points are also used in video editing, and we first used them in *Chapter 1*.

An In point is the point where we want our clip to start. The keyboard shortcut to create an In point is the *I* key, which is short for "In."

An Out point is where we want our clip to end. The keyboard shortcut to create an Out point is the *O* key, which is short for "Out."

In and Out points can be created for clips in the **Media Pool** to mark up the parts of clips to keep before they are added to the **Timeline**. Conversely, they can also be used to mark up part of a clip on the **Timeline** to be removed from the **Timeline**.

1. The start of the second clip that we have on the **Timeline** has both the bride and groom looking at each other, so we will need the end of our new clip in the **Media Pool** to also match this.

    Add an Out point (where the clip will end) by pressing the *O* key at 16:51:13:12.

2. Drag the clip down from the **Viewer** onto the **Timeline** just above Video Track 1. A new track (Video Track 2) will be automatically created, where we can drop our newly prepared B-roll clip (*Figure 8.4*):

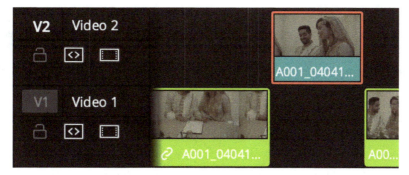

Figure 8.4: B-roll on Video Track 2

3. Position the clip on the **Timeline** in Track 2 so that it covers the missing clip on Track 1.

Now we have covered up the shaky camerawork at the start of the master shot by adding some B-roll to cover up the edit. The B-roll I have used is just a suggestion, so feel free to find your own part of the clip to use if you prefer a different angle.

> **Key concept – master shot**
>
> Although we have referred to the master shot in *Chapter 1*, we haven't needed to explain its use until now.
>
> A master shot is the main shot that tells our story, as it shows most of the scene in that one shot. Usually, a master shot is a wide shot, but it can be closer if the widest shot in the scene is a mid-shot, in which case, the mid shot will be the master shot.
>
> Other shots are then added to the master shot, such as cut-ins, cutaways, or B-roll to help add to the story.

The main learning point from this exercise is that when covering up a missing clip with a shot from another angle, we need to make sure that the action at the end of the previous clip matches the start of the new clip, and the end of the new clip matches the start of the next clip on the **Timeline**. If we do not do this, it will create a jump in the action, which will be jarring for the viewers. This principle of hiding obvious edits to help maintain the timing of the story is the basis of continuity editing.

> **Key concept – three-point editing**
>
> The process of adding an **In point** marker and an **Out point** marker to a clip, and then inserting them on the **Timeline** from a set point marked by the playhead, is called three-point editing.
>
> Once you have mastered the basics of drag and drop editing, as covered in this book, you can use your understanding of three-point editing to progress on to using the more advanced editing tools in the toolbar, such as inserting a clip – *F9* – or overwriting a clip – *F10*.
>
> Without going into too much detail (as that is beyond the scope of this book), you mark up your three points using In and Out points and your playhead, then use either the insert clip option (by pressing the icon or the *F9* key on your keyboard) or the overwrite clip option (by pressing the icon or the *F10* key on your keyboard). This will do the same job as drag and drop editing but more quickly, as you don't need to use your mouse.

This footage we have just added is not really a cut-in, as it is a shot of the same subject from a different camera angle. Now we are going to add a cut-in to help emphasize a certain part of the story.

Creating cut-ins

Here, we will use the same clip to show how cut-ins work. We need to add a close-up of the groom signing the marriage certificate as a cut-in to help tell the story. We want to emphasize this part of the story, and getting closer will draw our audience's attention to this key moment:

1. Move the playhead to the following timecode of the **Timeline**: 01:00:58:08.
2. Add a cut using the razor tool (the *CMD + B* key combination is the shortcut). This is the point where we will add our cut-in.

3. Now, let us mark up our cut-in on clip `A001_04041647_C034.mov`:

    A. Double-click on the clip in the **Media Pool** to load it in the **Viewer**.

    B. Move the playhead in the **Viewer** until `16:48:37:11`, and mark an In point (*I*).

    C. Move the playhead in the **Viewer** until `16:48:42:08`, and mark an Out point (*O*).

4. Drag the marked-up clip down from the **Viewer** onto Video Track 2 of the **Timeline**. Line up the start of our cut-in with the cut on Video Track 1 that we just made in *step 2*.

5. Play back the footage on the **Timeline** from the newly added cut-in.

    You will notice that the hand of the groom is on the page in the cut-in, then in the next shot, it is midair giving the pen to the bride. This jump has created a break in continuity that we need to fix.

6. Drag the third clip on Video Track 1 so that the start lines up with the end of the cut-in you have just placed on Video Track 2 in *step 4* (*Figure 8.5*):

Figure 8.5: Moving the master shot to line up with the end of the cut-in

7. Play back the clips on the **Timeline** to review your cut-in to see if the action flows. If it does not flow, you can trim the start and end points and reposition the clips on Tracks 1 and 2 until you get a closer match that you are happy with.

As you can see, adding cut-ins can require the back-and-forth trimming needed for continuity editing until it feels right.

Now, let us add a cutaway, which is usually a shot that does not affect the continuity of the scene but can reveal some internal thoughts the subject might be having.

Creating cutaways

When inserting in a cutaway, it is best to find a particular point in the main shot that motivates the cutaway. Play through the edit on the **Timeline** and mark any key moments where the subjects look away off-camera or seem to be thinking. Here's an example of how to do that:

1. Move the playhead to the following timecode of the **Timeline**: `01:01:21:03`. Both the bride and groom see something outside the shot (off camera) that attracts their attention. Notice that their eye-lines are looking to the lower-left part of the frame. We will need to find a B-roll clip where the action matches the direction of their eye-lines.

2. In the **Media Pool**, locate clip A001_04041629_C028.mov and load it into the **Viewer**.

3. Play it until you reach 16:29:09:02, where the B-Roll is of a little boy and girl walking up the steps. This shot matches the direction of your couple's eye-lines.

    Add an In point mark.

4. Add on Out point mark at 16:29:13:07 before the children reveal they are walking up to the couple in a different location.

5. Drag the marked-up clip down from the **Viewer** onto Video Track 2 of the **Timeline** (*Figure 8.6*). Line up the start of our cutaway with the **Timeline** playhead we just positioned in *step 1*:

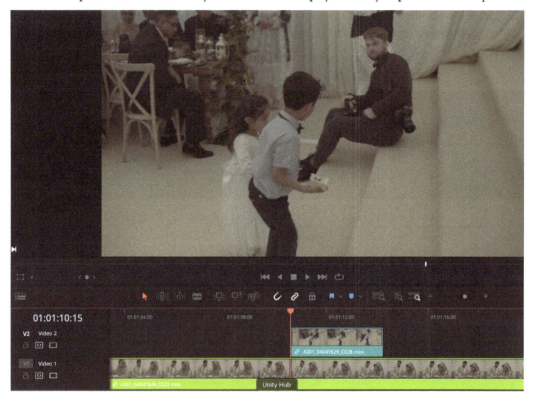

Figure 8.6: Cutaway lined up on the Timeline

6. Play back the footage on the **Timeline** from the newly added cutaway to review it.

    It now looks like the bride and groom are looking at the children approaching them.

Now that we have added both cut-ins and cutaways to our edit, let us look at other ways we can add an invisible edit.

As an optional exercise, see whether you can add an additional shot from B-roll to add some more visual interest. Once you have done this, compare your edit with the edit I have included in the **Wedding_Video Timeline**, which is part of the media download for *Chapter 7*.

Sometimes, there are occasions where we need to cut out a section of a clip but don't have any suitable B-roll footage to hide the edit. How do we then hide the resulting jump cut? Well, DaVinci Resolve has a solution to remove subtle jump cuts in the same clip – it is an effect called **Smooth Cut**.

Using Smooth Cut to hide an edit

Smooth Cut works best when there is not too much movement within the frame.

Much like the frame replacer FX plugin we used in *Chapter 6*, to replace a missing frame for the archive footage, the Smooth Cut video transition takes information from frames on both sides of the cut and blends them to hide the cut.

In this exercise, we are going to cut out the section of the clip where the groom looks at the camera, as this might be distracting for the audience and can break their immersion in the intimacy of the scene. Then, we will be removing the resulting jump cut with a Smooth Cut transition:

1. Move the playhead to timecode 01:00:53:12 of the **Timeline** and add a cut.

2. Move the playhead to timecode 01:00:56:19 of the **Timeline** and add a second cut.

3. Ripple delete this section of the clip by pressing the *Delete* key on your keyboard. The section of the clip will be deleted, and the rest of the clips on the **Timeline** will be rippled up to fill the gap.

4. Play back this new cut (see *Top tip – play around*) and you will notice a small jump cut, like a glitch in the Matrix. We will now fix this.

Top tip – play around (/)

A quick way to review a transition is to use the **play around** command, which is the / key on the keyboard.

The play around command will play the **Timeline** from 2 seconds before the playhead or what you have selected on the **Timeline** (either a clip or a transition) to 2 seconds after.

Play around is good for quickly reviewing edits you have made to see whether the edit has worked.

---

5. Open the **Effects** panel, and in the **Toolbox** section, navigate to **Video Transitions > Dissolve > Smooth Cut** (*Figure 8.7*):

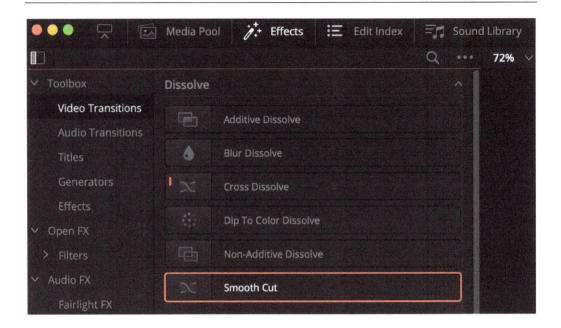

Figure 8.7: Smooth Cut in the Effects panel

6.   Drag the **Smooth Cut** transition onto the jump cut we created in *step 3* (*Figure 8.8*):

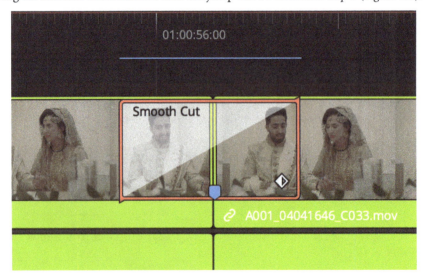

Figure 8.8: The Smooth Cut transition applied and rendered

7.   Play back the transition, and you will notice that the transition is very stuttery in playback. Let us make sure that the playback performance of any transition is optimal.

8.   To do that, in the menu bar, select **Playback** > **Render Cache** > **Smart**. This will mean that the transition will be automatically rendered when we play it back to help it play smoothly:

- A red bar on the **Timeline** above the transition means it needs to be rendered

- A blue bar (*Figure 8.8*) on the **Timeline** above the transition means that the effect has been rendered and will play back smoothly

Play back the transition, and you will notice that although it has improved, it is still a bit jumpy. The groom's face seems to have morphed (*Figure 8.9*). This is because Smooth Cut works best when there is little movement in the frame. In this clip, the groom moves his head, and the bride barely moves. You will notice that the glitch is at the start of the Smooth Cut, where it overlays the first clip. After that, the second clip plays back flawlessly. This gives us a clue on how to fix it:

Figure 8.9: The groom's face showing an unwanted morph effect

We can fix this by changing the **Smooth Cut** settings.

9.   Select the **Smooth Cut** transition (an orange border will appear to show it has been selected; *Figure 8.8*).

10.   Open the **Inspector** panel to reveal the transition controls (*Figure 8.10*).

11.   Change the **Transition** settings until you get a smooth result. These are the settings I chose for my **Smooth Cut** transition for this edit:

Figure 8.10: The Smooth Cut controls in the Inspector panel

A.   **Duration > Seconds and Frames**: By default, this will be *1* second or 25 frames, as this clip has a frame rate of 25 **frames per second** (25 **fps**). You can change the duration of the clip by changing either the **Seconds** or **Frames** value. I am going to change the transition to 9 frames (0 . 4 seconds) long as a shorter Smooth Cut often produces better results.

B.   **Alignment**: There are options here to choose whether to apply the Smooth Cut only to the clip to the left of a cut (the left-hand button), over the clips on both sides of the cut (the middle button), or just the clip to the right of the cut (the right-hand button

I chose the right alignment, as, if you remember, the glitch was to the left of the cut, so we can avoid this glitch altogether

C.   **Mode**: This drop-down menu selects the quality of the render:

    i.   **Faster** renders faster but with a slight drop in quality.

    ii.   **Better** is the highest quality render but takes longer to do so

D. I have left this at the default **Better** setting.

E. **Ease**: This drop-down menu selects the following:

- **None**: This is the default setting that applies a linear transition

- **In**: This eases in the start of the transition

- **Out**: This eases in the end of the transition

- **In & Out**: This eases in the start and the end of the transition

- **Custom**: This is where you can add your own keyframes to the transition curve to modify how gently or suddenly you want the transition to be applied

- I have left this at the default **None** setting

12. Now you can play back the transition on the **Timeline**.

If it is still a bit jittery, you can change the settings in *step 11* to your own preference until it looks right to you.

Smooth Cuts are great for removing jump cuts that appear when we edit out all those "umms" and "aahs" in our online videos.

The next way to disguise a cut is not with any additional footage but to use audio to smooth over the transition. This is called a **Split Edit**.

Hiding cuts with a split edit

A split edit is a cut where the transition between the audio and video is split to happen at different times.

There are two types of split edit, a *J cut* where we hear the audio from the next clip before we see the video, and an *L cut* where the audio from the previous clip is heard over the video of the next clip. They are named *J* and *L* cuts after the shape they make on the **Timeline**. The bottom of the *J* and *L* cuts is the audio, and the upright shape of the letter is the video, as the audio appears below the video on the **Timeline**.

Split edits are good for hiding an edit, as there is a softer transition than a straight cut (through cut), and our brains only have to deal with one sensory change at a time. With a straight cut, the video and audio change at the same time, whereas with a *J* or *L* cut, the audio changes at a different time in the video, which softens the edit and makes it less obvious.

With this exercise, we need some audio to play with, and as the wedding video does not have any good audio, I have created an archive file especially for this exercise.

You can download the archive file here:

https://packt.link/B5bqz

We are going to use the **Edit** page **Timeline** to demonstrate how a split edit works:

1. Open the archive file called **Split Edit Explainer** on the **Edit** page.

    You will notice that all the clips are already on the **Timeline**, with cuts already applied. Now we are going to add some split edits.

2. First, we need to turn off the linking of the audio and video by clicking on the **Link** button (which looks like chain links) just above the **Timeline** (*Figure 8.11*):

Figure 8.11: The Link/Unlink button (the second white button from the left-hand side)

    Turning off the **Link** function enables us to now move the edit point for the audio separately to the video.

3. Under the second clip, where I describe a *J* cut, drag the edge of the audio to the left-hand side so that it comes under the video title clip. Move the audio until the phrase *a J cut is where we hear the audio of a clip before we…* is underneath the title clip.

4. Play back the edit. When you play this back, you will hear me describing what a J cut is before you see me talking.

5. After the third clip, where I describe an *L* cut, drag the edge of the video from the end title to the left-hand side so that it covers over some of the third clip. Move the video until the phrase *whilst we still hear the audio from the previous clip* is heard underneath the end title clip.

6. Play back the edit. When you play this back, you will continue to hear me describe what an *L* cut is after the video has cut to the next clip.

7. Once you have finished, remember to re-enable the **Link** button to stop you from accidentally making split edits when moving clips on the **Timeline**.

That's it; we have now made our first split edits. It is best to use a Split Edit cut if there is a motivation to do so in the scene; otherwise, it can come across as a mistake.

Split edits are great to show a back-and-forth conversation, as real-life conversations usually overlap.

For instance, let us say that I am watching two people, person A and person B, have a conversation. I am looking at person B when I hear person A talking to person B. I hear person A speak first (while I am looking at person B). Then, I turn to see person A speaking, pretty much just like a J cut. If I want to see person B's reaction to what is being said, I will look at them while person A is still talking, just like an L cut. In this situation, we would be showing the audience that we are interested in the emotional journey of person B during the conversation.

We have now used a variety of editing techniques to help the flow of a story and to hide our editing. This brings us to the end of this chapter.

Summary

Here is what you have achieved in *Chapter 8*:

- Learned what cutaways and cut-ins are and why they are used
- Learned what continuity editing is
- Learned how to create a cut-in on the **Edit** page
- Learned how to create a cutaway on the **Edit** page
- Used a Smooth Cut transition to hide a jump cut
- Learned what a split edit is (J and L cuts) and how to create them

Learning how to edit using B-roll and split edits gives us a greater opportunity to tell a seamless story where the viewer is more immersed in our story and not distracted by our editing techniques.

In *Part 3* of the book, *Advanced Techniques*, we start with *Chapter 9*, where we will look at how to shoot for a green screen so that we can change our backgrounds for our videos in DaVinci Resolve.

Questions

1. True or false? A cutaway or cut-in is another name for a split edit.
2. True or false? Continuity editing is when we add the audio of a continuity announcer.
3. True or false? A cut-in is used to emphasize a closer detail within the action.
4. True or false? A cutaway is when we remove unnecessary video footage.
5. True or false? A Smooth Cut is a type of transition.
6. True or false? There are two types of split edits, a J cut and an L cut.

Further reading

Here is a list of some useful articles on editing theory, such as cutaways and cut-ins, to help add a deeper understanding of what we have covered in this chapter:

- `https://www.masterclass.com/articles/cutaway-shot-explained`
- `https://www.studiobinder.com/blog/cutaway-shot/`
- `https://nofilmschool.com/cutaway-shot-definition-examples`

Here is a video demonstrating a split edit:

- `https://www.youtube.com/watch?v=9tWpSl5co_Q`

Part 3:
Advanced Techniques

The objective of this section is to familiarize you with more advanced techniques that can further enhance your footage. We will do this by enhancing colors, applying visual effects, creating green-screen or split-screen effects, and using Studio-only, AI-based tools.

This section comprises the following chapters:

- *Chapter 9, Adding Special Effects*
- *Chapter 10, Split Screens and Picture-in-Picture*
- *Chapter 11, Enhancing Color for Mood or Style*
- *Chapter 12, Studio-Only Techniques*
- *Chapter 13, Glossary*

9
Adding Special Effects

A green screen has typically been used for **visual effects** (**VFX**) in the film industry such as to replace a background to show a superhero flying, or on television to show a weather forecaster's map. However, there is no reason why you cannot use it to spice up your videos for social media or the web.

The need for using green screens on personal videos may have various motivations. For example, sometimes you may want to have the option to change the background of a video after it has been filmed, such as adding an animated graphic (for example, showing key business stats) in the background behind your presenter.

In the same vein, it is also good to use a green screen when you want to update the background of your video to suit different markets while keeping the foreground content the same.

In this chapter, we will cover how to shoot using a green screen and how to work with greenscreen footage in DaVinci Resolve.

In this chapter, we are going to cover the following main topics:

- Understanding codecs
- Shooting video for greenscreen
- Chroma keying in DaVinci Resolve
- Creating a greenscreen effect on the **Edit** page using the **3D Keyer**
- Creating a greenscreen effect on the **Edit** page using the **HSL Keyer**
- Removing chroma key background spill using the **Despill Resolve FX** plugin

Technical requirements

We will be practicing our greenscreen skills using footage by artist Stefania Buzatu (*StefWithAnF*), found on *Pixabay*: https://pixabay.com/users/stefwithanf-795875/?tab=videos&pagi=8.

To follow along with the exercises in this chapter, download the following video, which contains all the greenscreen clips needed: https://packt.link/B5bqz

Or you can use your own greenscreen footage instead. However, to get good results when using a green screen, it is necessary to know the best file type or codec to use when shooting video for greenscreen.

Understanding codecs

Usually, when we save our video in any editing software, including Resolve, we will be presented with a confusing array of different file formats to save our video in.

The main difference between each of these video file formats is the codec that they use. When exporting our video on Resolve's **Deliver** page, we are shown options for both **Format** and **Codec** (*Figure 9.1*):

Figure 9.1: Export Video settings on the Deliver page

Essentially, a video file format such as .MOV or .MP4 (shown by a dot followed by a three- or four-letter code at the end of a filename) is a container to store the video file in.

Each of these video files uses different ways to squeeze the video and audio information into the container. The way the video is squeezed into this file container is called a codec.

What is a codec?

A codec is a set of instructions (known as code) for the camera to help it compress a video to store in a computer file and then uncompress it on the computer so that you can view it again. In fact, *codec* is a short way of saying *compression decompression* or *code decode*.

To understand a codec in a practical sense, let's use the example of an inflatable mattress. To store it, you need to remove all the air from it by rolling it up and then compressing it into a bag. However, to be able to sleep on your mattress, you need to remove the mattress from the bag and reinflate it with replacement air. The process of deflating your mattress and reinflating it is what a codec does to a video file, except that instead of air, it removes video data.

Types of compression

How the codec removes data and restores it differs for each codec. Video files can be compressed in two different ways: **temporal compression** and **chroma subsampling**.

Temporal compression

Temporal is the Latin word for time. This compression treats each frame as a picture and looks for how they change over time. Not every aspect of the frame will change. This type of compression is where the codec identifies key frames that are kept and deletes the frames in between the key frames. Basically, the differences in the visual data from one key frame to another are used to recreate the missing frames in between the key frames when the video file is played back.

Temporal compression works well when there are slow changes between the frames with very little movement and with large blocks of color, such as a blue sky, as it is easier to recreate the lost visual data. Where temporal compression does not work so well is in fast-moving scenes with a lot of detail where there are rapid changes, as this detail is not easy to recreate accurately.

Files that heavily rely on temporal compression can be difficult to play back in real time as the computer has to recreate the missing frames as it plays the video file.

Common types of temporal compression are H.264 and H.265.

Chroma subsampling

Chroma comes from the Greek word for color. This type of compression deletes color information from every other pixel of the original video file and copies (samples) the color information from the adjacent pixels when it is played back.

Chroma subsampling takes an image and divides it into blocks of 4 x 2 pixels. Subsampling is applied to each of these 4 x 2-pixel blocks. Which pixels to keep and sample the color information from is based on the subsampling ratio. Chroma subsampling is described as a ratio of three or four numbers—that is, 4 : 4 : 4 or 4 : 4 : 4 : 4.

The extra number at the end of the 4 : 4 : 4 : 4 ratio is based on the video having an **Alpha (transparent) channel**, which is mainly used for VFX or composite work where it is useful to have a transparent background—this is beyond the scope of this book.

For this book, we will be looking at the three-number ratio. The first number in the ratio is based on the width of the pixel block and is always 4. The second number in the ratio is the number of samples across each row of pixels. The third number is the number of color-sample changes in the second row of pixels compared to the first row. Generally, with the third number in the ratio, the color-sample changes are either the same as the first row or zero. Here are some examples of what this might look like.

Figure 9.2 shows the original image with no subsampling applied. Eight colors are shown. This is the full-color resolution:

Figure 9.2: 4:4:4

Figure 9.3 shows subsampling applied where all four colors are kept from the first row and there are zero changes between rows one and two, hence the number 0 at the end of the ratio. So, only four colors are shown. This is half the original's color resolution (that is, it now only stores four colors instead of the original eight colors per sample, making the image size much smaller):

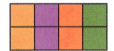

Figure 9.3: 4:4:0

Figure 9.4 shows subsampling applied with only two colors kept from the first row, and there are two changes between rows one and two. In total, four colors are shown. This is half the original's color resolution. This is the most used subsampling ratio as it is a good blend of color fidelity and compression and is much better looking than 4 : 4 : 0 with the same amount of compression:

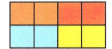

Figure 9.4: 4:2:2

Figure 9.5 shows subsampling applied with only two colors kept from the first row, and there are zero changes between rows one and two. Two colors are shown. This is a quarter of the original's color resolution. Most consumer video cameras will create video files using this ratio:

Figure 9.5: 4:2:0

Figure 9.6 shows subsampling applied with only one color kept from the first row, and there is one change between rows one and two. Two colors are shown. This is a quarter of the original's color resolution. This ratio is hardly ever used as 4 : 2 : 0 looks much better than 4 : 1 : 1 with the same loss of color:

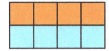

Figure 9.6: 4:1:1

Figure 9.7 shows subsampling applied with only one color kept from the first row, and there are zero changes between rows one and two. Only one color is shown. This is one-eighth of the original's color resolution. This subsampling ratio is very rarely used as there is such a loss of quality:

Figure 9.7: 4:1:0

As you can see, choosing the right chroma subsampling ratio can help retain the correct color information. Some file format codecs allow you to choose the chroma subsampling to apply. *Figure 9.8* shows the **Render** settings on the **Deliver** page with Apple's **QuickTime** file format with the **Apple ProRes** codec applied. With the Apple ProRes codec selected, you can choose a chroma subsampling ratio of **4444** (with no loss of quality) or **422** (with half of the color information). **Apple ProRes 4444** will be a much larger file size than **Apple ProRes 422**:

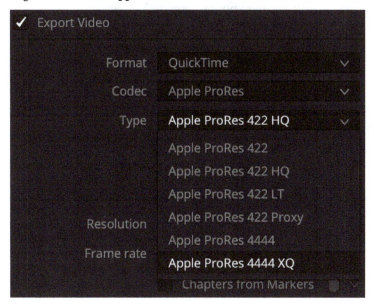

Figure 9.8: Chroma subsampling options when exporting video

Now, let us look at the different situations where we can use this knowledge of compression types to get the most out of our video images.

Types and uses of codecs

What type of codec compression is used for your video file depends upon how you are using the video.

Acquisition codecs

These are the codecs used by your video camera to record a video file. These are based on a balance of storage space and being quick to write to your storage medium, such as an SD card.

Some of the common consumer camera codecs are as follows:

- MPEG-4 (.MP4), QuickTime (.MOV), and AVCHD (Sony and Panasonic cameras) video file formats all use H.264 as a codec.

- Newer consumer cameras, including iPhones, give the option (HEVC file format) to use H.265 as a codec, which is a newer, more efficient version of H.264. H.265 also supports 8K video files, whereas H.264 doesn't. However, as H.265 is a newer format, it is not supported for playback on everyone's devices.

> **Key concept – what about RAW files?**
>
> Camera RAW files are neither codecs nor file formats. They are the original uncompressed raw data captured by the camera before it has been compressed into a file format. Think of RAW as being a digital negative of which you can change the exposure of the video afterward.
>
> Each camera manufacturer creates professional and semi-professional cameras that record their own camera data into a RAW file—for example, BRAW is Blackmagic Design's raw camera format. To be able to see the recorded video, you need to convert it into a common video format first or use a dedicated video player app (for example, Blackmagic RAW Player) that converts the RAW video as it plays it back.

Intermediary codecs (proxy media)

Intermediary codecs are codecs that are optimized to ensure that the playback on your computer when editing a video file is quicker and smoother. Intermediary codecs are also called proxy media as they are used in proxy (replacement) for the original video camera media.

Because spatial compression codecs such as H.264 and H.265 have missing video frames that need to be recreated as they play back, they are not great for editing with. When editing a video, you often need to stop on a particular frame and play forward and backward quickly as you make changes. This can cause the computer to slow down as it calculates the missing video frames to show you in the video viewer as you play it back. The video in the **Viewer** can then appear slow and out of sync with the audio, which makes editing a bit of a chore.

For this reason, video editors use what is called an intermediary codec, which is easier to edit with. The original camera format is converted into a codec that does not use as much compression and is hence easier to play back and edit.

There are two main intermediary codecs used in video editing—they are Apple's *ProRes* and Avid's *HNxHD* or *HNxHR*. HNxHR is Avid's newer intermediary codec that supports resolutions higher than HD 1080p.

To be able to create Apple's ProRes codec, you will need to own an Apple Mac, whereas Avid's HNxHD and HNxHR are supported on both Macs and PCs.

In DaVinci Resolve, intermediary codecs can be selected for use in your project in **Project Settings > Master Settings > Optimized Media and Render Cache** (*Figure 9.9*).

As a reminder from previous chapters where we have looked at **Project Settings**, you can find it by clicking on the little gear icon at the bottom right-hand corner of the screen:

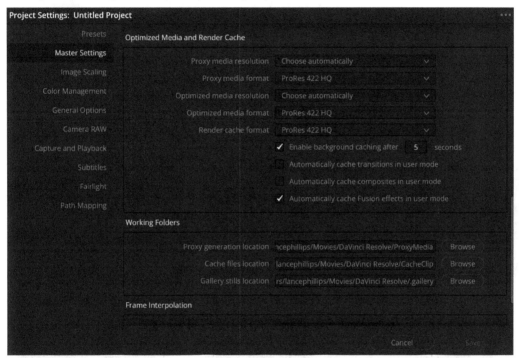

Figure 9.9: Project Settings showing proxy and optimized media

Resolve creates intermediary codecs in two ways: **Optimized Media** and **Proxy Media**.

Optimized Media is where Resolve creates the intermediary codec internally in DaVinci Resolve's media cache and does not save it as discrete media files on your computer. This is the quickest option.

To enable **Optimized Media**, follow these steps:

1. Create optimized media by right-clicking any clip or one of a group of selected clips in the **Media Pool** and selecting **Generate Optimized Media** in the pop-up menu.

2. To enable optimized media to be used: in the **Playback** menu, select **Use Optimized Media if Available**.

Proxy Media is where Resolve creates additional intermediary codec files outside of Resolve that you save to your computer's hard drive. This option is good if you want to share the proxy files with someone else so that they can work on them separately or remotely without having to spend the time to recreate intermediary codecs.

To enable **Proxy Media**, follow these steps:

1. Create proxy media by right-clicking any clip or one of a group of selected clips in the **Media Pool** and selecting **Generate Proxy Media** in the pop-up menu.

2. To enable proxy media to be used, see the following:

    *   **Resolve 17**: In the **Playback** menu, select **Use Proxy Media if Available**
    *   **Resolve 18**: In the **Playback** menu, select **Proxy Handling** > **Prefer Proxies**

You can select which codec to use for either optimized or proxy media in **Project Settings** > **Master Settings** > **Optimized Media and Render Cache** (*Figure 9.10*). Here, you can choose the format, resolution, and location of where the intermediary codecs will be saved:

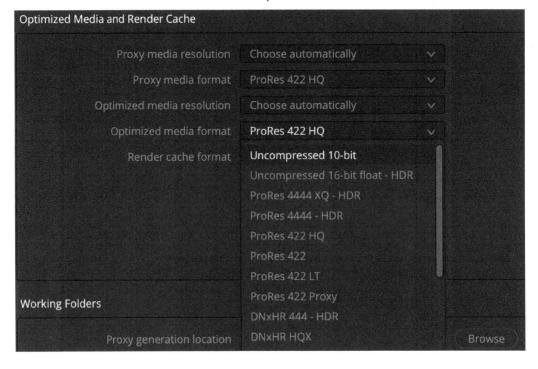

Figure 9.10: Project Settings showing the different intermediary codecs to choose from

Resolve does not limit you to choosing either **Optimized Media** or **Proxy Media**—you can use both in the same project.

> **Top tip – editing versus color grading codecs**
>
> If you are color grading, you will need to have the highest quality codec for color. So, choosing a codec that supports 4 : 4 : 4 : 4 is important as there will be no loss of color.
>
> You can easily select **Proxy Handling** > **Prefer Camera Originals** in the **Playback** menu, which will force Resolve to use the original camera files. You can also change the cache media codec to suityour purposes by selecting a 4 : 4 : 4 : 4 codec in **Project Settings** > **Master Settings** > **Optimized Media and Render Cache** (*Figure 9.10*).

Delivery codecs

These are codecs to make it easier (that is, no video lag) for the final video to be watched by the audience, depending upon the platform they choose to watch your video on, such as the web, a streaming service, or a TV broadcast.

Common delivery codecs for social media are QuickTime (.MOV) and MP4 in either H.264 or H.265.

There are presets in Resolve that select the best codecs for YouTube, Vimeo, and Twitter. As shown in earlier chapters, these can be found in the **Quick Export** menu.

> **Top tip – best codec for effects work**
>
> One thing to remember is that when doing any effects work such as greenscreen or applying **Fusion** or **Color** effects, the effects will need as much of the original footage quality as possible to get the best result. In this case, use a high-quality proxy compression format such as 4 : 4 : 4 : 4 or the original source files to get the best result.

Not only is it important to know which file type to use to shoot greenscreen but also how to shoot it to get the best results. This makes it easier for you to separate the background from your subject in your software editor.

Shooting video for greenscreen

Just to clarify, the process of removing the background of a video to replace it in post-production with a new background is commonly called **greenscreen**, which is not to be confused with the name of the background which is filmed as part of this process which is a green screen.

Let us explain how this greenscreen process works so that we can understand how to light our video especially for it.

We start by filming our central subject against a single-color background (usually a green screen). Then, in the editing software, we identify the color of the background we want to remove from the video. The software removes the specified color from the video, leaving behind a transparent layer in its place. This leads to a transparent background. We can then put another video or computer-generated graphics behind this transparent layer of our main video. The newly added video or graphics will be seen where the transparency has been created by removing the single color from our video in front.

Notice how I said a *single color*. Well, you can use any color for this process; it does not need to be green.

DaVinci Resolve does not use the term *greenscreen*. Instead, it uses the term *chroma key*. Let us explain what a chroma key is.

What is a chroma key?

Originally, in the days before digital videos were a thing, a blue screen was used (instead of a green screen) as this was the easiest color to remove from film. Now, with most video being digital, we use a green screen as it is easier to remove green, as the digital video signal has twice as much information about the different tones of green than red or blue.

Because any color can be used for the background screen, the technical term for the process of removing the background in videos is called **chroma keying**. *Chroma* is the Greek word for color (*khrōma*), while *key* is a Middle German word (*ki*) for splitting. So, chroma keying is splitting the video into two parts based on color. The background layer that is created is often called the *key*, and the process of chroma keying is often abbreviated to *keying*.

The most common colors used to split a video into an opaque foreground and a transparent background layer are green or blue. You can use any color if the background color is not featured in the foreground subject of the video. As a use case, when filming vegetables or plants as a subject, you would want to use blue as a background instead of green; otherwise, you would have parts of the vegetables become transparent in the video.

Here are some other top tips when filming for chroma keying.

Lighting your background

Light the background as evenly as possible so that there are minimum variations in the exposure of the background.

When lighting the background, we want to eliminate shadows in our background color as shadows have less color in them and are harder to remove later.

To be able to do this, use diffused (not hard) lights that are directed at the background and have a wide, even spread of light.

If any parts of your background will not be seen by the camera, block them off to minimize the chance of light from the background spilling (back-spill) onto the foreground subject.

Camera settings

Do not use your camera's ISO to boost the exposure of your video, as this introduces digital noise that can affect the quality of your *key*.

Do not use any lens diffusion or smoke effects, as these can soften the image and make it harder to recognize the boundaries between the subject and the background. For the same reason, make sure your subject is in focus and is as sharp as possible, so use the greatest **depth of field** (**DOF**) available.

Set your camera's white balance for the light on your foreground subject.

Include a gray card at the start of your video; getting your subject to hold the card is a good idea. Use the same lighting for the gray card as you are using for your foreground subject, making sure you do not get any back-spill onto the card or your subject. This gray card can be used by Resolve to help when making automatic color corrections, as described in *Chapter 11*.

Key concept – gray card

A gray card is a middle-gray-colored card that is filmed at the start of the shot to give a reference to editing software for what middle gray should look like. This helps when trying to color balance the video later in a post (that is, make sure the image is neutral and neither too warm nor cold in color).

Gray cards can also be used as a reference when manually setting your camera's white balance.

When using a zoom lens, do not use it at its widest setting as this can create a slight vignette, which will create problems for the background key.

Use the highest resolution and least compressed video codecs possible. If it has the option, use your camera's chroma-subsampling compression of at least 4 : 2 : 2; ideally, use a non-compressed codec using 4 : 4 : 4.

Filming your subject

One of the problems with lighting a single-color background is that the color from the background can spill onto our foreground subject; this is called back-spill. This presents a problem later as this will then remove the parts of the foreground subject where the background light has reflected onto them.

To eliminate back-spill, you need to keep the foreground subject as far away from the background as possible.

When lighting your subject, try not to create shadows from your foreground subject onto your background—that is, use a soft diffused light. This is also the case if you are using a chroma key for the floor. Using reflectors can decrease the shadows that spill on the floor.

It is best not to have your subject lying or sitting on a chroma-key floor or chair as this will create more opportunities for back-spill, and (if your subject is a soft-edged irregular shape, such as a person) it will be difficult to separate the foreground from the background. If you need to have your subject sitting or lying down, use a section of real floor, with defined edges, which can be merged with the background later.

Make sure your subject does not have any colors that are in your background. For example, when using a blue screen, make sure your subject is not wearing blue jeans or a blue shirt.

Now that we know how to get the best-quality footage for our green screen, let us look at how we can create a greenscreen effect in DaVinci Resolve.

Chroma keying in DaVinci Resolve

For the following exercise, we will remove the background a of woman reading George Orwell's *Nineteen Eighty-Four* and replace it with an infinite zoom video showing dystopian settings, reflecting what she is reading about in the book.

Preparing our Timeline for the key

Even though we can apply Resolve's **OpenFX Keyer** plugins to clips on the **Timeline** on the **Cut** page, we do not have the ability to select the color we want to remove (key out).

You can import the video clips and get the **Timeline** ready for keying on either the **Cut** or **Edit** page. Here, we describe the **Cut** page. If you want to use the **Edit** page, you can apply the same principles. Proceed as follows:

1. In the **Cut** page's **Media Pool**, select `Infinite Zoom - 44658.mp4` and put it on **Video Track 1 (V1)** of the **Timeline**.

    This will be our background clip. It is important to always put out a background clip on the lowest video track, usually **V1**.

2. In the **Cut** page's **Media Pool**, select `Reading - 40098.mp4` and put it on **Video Track 2 (V2)** of the **Timeline**.

    This will be our foreground clip. It is important to always put out a foreground clip on the highest video track, usually **V2**.

3. Trim the end of `Infinite Zoom - 44658.mp4` on **V1** to be the same length as the `Reading - 40098.mp4` clip on **V2**:

    This is not necessary for the keying process but does make our **Timeline** a bit tidier.

That's it—we are now ready to apply our keyer to the topmost clip on the **Timeline**, after which we will remove the green, revealing the background clip underneath.

It can be confusing to decide which keyer to use in DaVinci Resolve as there are several options. In this chapter, we will look at the main two chroma keyers available on the **Edit** page of Resolve: the **3D Keyer** and the **HSL Keyer**.

Creating a greenscreen effect on the Edit page using the 3D Keyer

The **3D Keyer** is a chroma key tool where you can quickly draw lines over the color you want to key (remove) from the background.

Every line you draw selects the corresponding colors in the 3D color space of the video. A **color space** is a 3D representation of all the colors available for your video. See *Chapter 11*, for a more detailed explanation of color spaces. Hence why the tool is called the **3D Keyer**, as it takes a key from the colors in this 3D color space.

You can apply the **3D Keyer** on either the **Cut** or **Edit** page (of course, you can also use the **Fusion** and **Color** pages, but that is a bit more complicated than you need it to be).

Applying the 3D Keyer

After locating the **3D Keyer** in **Effects** > **Video** > **Resolve FX Key** > **3D Keyer**, you can apply the **3D Keyer** to the selected clip in the **Timeline** using any of the following methods on either the **Cut** or **Edit** pages (*Figure 9.11*):

- Double-click on the **3D Keyer** in the **Effects** tab
- Drag and drop the **3D Keyer** from the **Effects** tab to the selected clip in the **Timeline**
- **Cut** page only: Click the **Add Effect** button (located at the bottom of the **Effects** tab)

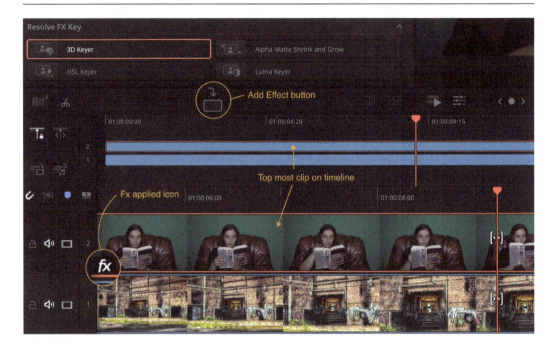

Figure 9.11: Applying the keyer on the Cut page

Make sure that you have selected the topmost video clip on the **Timeline** before applying the keyer; otherwise, you will not see the effect as it is hidden under other video tracks.

You will notice that a small **fx** icon (*Figure 9.11*) has appeared on the bottom left of the clip on the **Timeline**. This is a quick visual reminder that this clip has an effect applied to it.

Adjusting the 3D Keyer

Now that we have applied our **3D Keyer**, we need to adjust it to remove (key out) the green background to reveal our video track underneath. Proceed as follows:

1. On the **Edit** page, open up the **Inspector** to reveal the keyer's controls in the **Effects** panel (**Inspector** > **Effects** > **3D Keyer**).

2. We now need to activate these controls in the **Viewer** by selecting **Open FX Overlay** in the **Viewer** overlay menu at the bottom left-hand side of the **Viewer** (*Figure 9.12*):

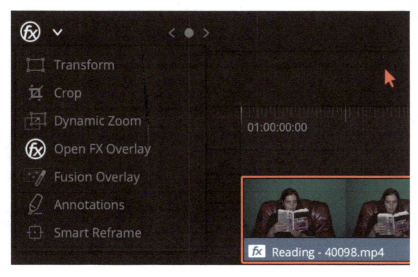

Figure 9.12: Open FX Overlay

3. Clicking in the **Viewer** on any part of the green background will remove most of the green to reveal the video image underneath, although it is a good idea to draw a stroke on the green background close to the subject, as this is where there will be the most variance of green to be able to get a good initial selection. Wherever you first select, we will still need to tidy this up as there are still parts of the green background showing.

4. We need to enable some options so that we can see what we are doing when selecting more parts of the green to remove:

   I.   Under **Usage Options**, make sure **Show Paths** is selected (checked).

   II.   Under **Output**, select **Alpha Highlight B/W**.

   You will now notice two things (*Figure 9.13*):

   • You can now see blue lines that show where you have clicked or dragged your mouse on the **Viewer** to key out the background.

   • The image has now turned black and white (**B/W**). The white shows where the image is opaque (solid), and the black shows where the image is transparent (see-through).

   Our task is to turn the entire green background black while making sure our subject in the foreground stays white:

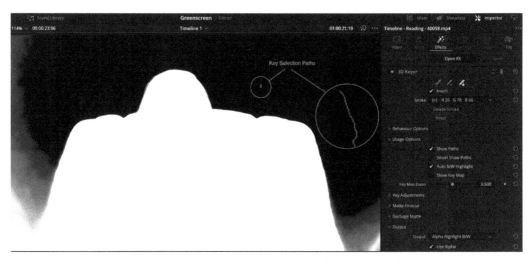

Figure 9.13: Viewer showing Alpha Highlight B/W and paths

5. Select the **Add** tool (eyedropper icon with a plus (+) symbol next to it). This tool will allow us to add to our initial selection.

6. With your mouse, click and drag on the white parts of the **Viewer** to select them to turn them black (and hence transparent).

    If you make a mistake, you can remove the stroke you just added by selecting the **Delete Stroke** button or use the **Subtract** eyedropper tool (with a minus (-) symbol next to it) to select black parts to become white again.

7. Once you have removed most of the background, you can check your selection by changing your **Output** value to **Final Composite** to see the finished result.

    You may notice that the only green that remains is around the edges of the subject. This is tricky to select without removing the subject as well. Well, we have a tool to help with this task.

8. Select **Alpha Highlight** in the **Output** menu. This will make it easier to see the green edges as it reveals the foreground subject while showing the removed background as gray (*Figure 9.14*):

Figure 9.14: Viewer showing Alpha Highlight and a green fringe around our subject

9. To remove this green fringe, reveal the controls in the **Inspector** under **Matte Finesse**. This is where we can fine-tune the section of the background we want to remove. The area of the background we want to remove is called **Matte**. Proceed as follows:

    - Move the **Clean Black** slider to the right to expand the amount of the background selected.

    - You can also move the **Black Clip** slider to the right to refine the edges of the background selected.

    - The **In/Out Ratio** slider will refine the selection by changing where the black and white selections meet. Move it to the left, and it shrinks the white and expands the black. Move the slider to the right, and it has the opposite effect.

    - Finally, we can increase the **Blur Radius** slider to soften the edges of our selection between the foreground and background to make it less obvious and blend the two together.

    You may have noticed a slight green tint around the edge of the subject's hair. This is caused by some light reflected off the green screen and spilling onto our subject. This green tint looks nasty and draws our attention to the fact that this video has been greenscreened. So, next, we'll take care of this.

10. To remove the green lighting spill from the background, move the **Despill** slider, in the **Behaviour Options** section, to the right until the green spill disappears.

11. Other tools you can use to refine the key are the following:

    - The **Pre-Filter** slider (on *page 1* of the **Matte Finesse** tools) cleans up the video image before the colors are keyed. This is a good tool to use if your original footage is low-quality and you need to tidy it up to help get a successful key.

    - You can also use **Post-Filter** (on *page 2* of the **Matte Finesse** tools) to bring back some detail lost from the edges after the key (particularly the hair).

12. Now, it is just a matter of tweaking the **Matte** area by adjusting the preceding controls and using the **Final Composite Output** setting to check the result as you go.

You now have a keyed-out background that reveals the background video behind. Play back the video to check your work (*Figure 9.15*):

Figure 9.15: The finished result

> **Key concept – matte**
>
> I know it sounds like the name of one of your best friends. However, *matte* is an old French word that means both the backing of a picture and a dull colorless surface. If you think of it, this is a good word to use to describe removing a color to reveal a background behind a picture.
>
> In early cinema, before we had computer graphics, matte artists or matte painters would paint a **background matte** onto glass to recreate a background that was impossible to film in real life (especially useful in sci-fi films such as the original *Star Wars*). The camera would then film the real foreground through the glass of this matte painting to create the illusion that the whole scene was filmed as one. The matte painting on glass and the foreground subject being filmed were carefully lined up so that you could not see the join. See the *Further reading* section of this chapter for links to a more detailed description of this process.
>
> Now, in video compositing software, we refer to the background we have selected to remove from the video as the matte.

Even though the **3D Keyer** is a quick and easy way to get a successful key, there will be times it will not work because of the nature of the video. If there is an image that needs more control to get a successful key, we can use the **HSL Keyer**.

Creating a greenscreen effect on the Edit page using the HSL Keyer

The **HSL Keyer** has different options from the **3D Keyer** as you can choose to select the key based on the *hue*, *saturation*, or *luminance* (*HSL*) of pixels in the image. This allows you to have more control over trickier keys as you can select any combination of hue, saturation, or luminance to get a successful key.

For example, there may be a background that is not a uniform green that also matches the green the subject may be wearing. However, if the background is less saturated and bright than the subject, you can still separate the background by only selecting the saturation and luminance of the green background and not the hue.

However, unlike the **3D Keyer**, the **HSL Keyer** does not have a **Despill** option, so it is best to only use it if you cannot get a successful key using the **3D Keyer.**

Let us create a new **Timeline** to try the **HSL Keyer** on different footage. To get our greenscreen footage and create a new **Timeline**, follow these steps on the **Edit** page:

1. In the **Media Pool**, open the bin named **HSL Keyer**.
2. Right-click the clip named HSL_MCU.mp4 and select **Create New Timeline Using Selected Clips…**.
3. In the **Timeline Name** dialog box, name your **Timeline** HSL_Keyer.
4. Click the **Create** button.

You now have a new **Timeline** with our greenscreen footage ready for us to apply our **HSL Keyer**.

You can either add the background video track now—as we did with a **3D Keyer**—or add it later. This time, I am going to add the background after I have keyed the greenscreen clip.

Let us apply the **HSL Keyer** now.

Applying the HSL Keyer

Just as with the **3D Keyer**, you can apply the **HSL Keyer** on either the **Cut** or **Edit** page.

After locating the **HSL Keyer** in **Effects** > **Video** > **Resolve FX Key** > **HSL Keyer**, you can apply the **HSL Keyer** using any of the same methods described for the **3D Keyer**.

Adjusting the HSL Keyer

Now that you have applied the **HSL Keyer**, you can change the image qualities (**Hue, Saturation, Luminance**) that Resolve selects to get a key. Trying different combinations of keying using **Hue, Saturation**, and/or **Luminance** can solve difficult keys, as any one of these image qualities could create difficulties for the key. Proceed as follows:

1. On the **Edit** page, open the **Inspector** to reveal the **keyer's** controls in the **Effects** panel (**Inspector** > **Effects** > **HSL Keyer**).

2. Activate these controls in the **Viewer** (just as you did for the **3D Keyer**) by selecting **Open FX Overlay** in the **Viewer** overlay menu at the bottom left-hand side of the **Viewer** (*Figure 9.12*).

3. Under **Keyer Options**, you can select how Resolve chooses which part of the image to key, whether it is based on the hue (color), saturation (strength of color), or luminance (how bright the image is). You can turn these options off and on individually by selecting the checkbox next to each option (*Figure 9.16*).

4. It is a good idea to turn each one on and off to see how it affects the key, then use the combination that works best for the clip you are keying. With this clip, I have chosen to leave all three enabled: **Use Hue, Use Saturation**, and **Use Luminance** (*Figure 9.16*).

5. Just as when we used the **3D Keyer**, clicking in the **Viewer** on any part of the green background (using the **Pick** eyedropper to make the initial selection) will remove most of the green to reveal the video track underneath. However, with the **HSL Keyer**, we cannot see the selection strokes.

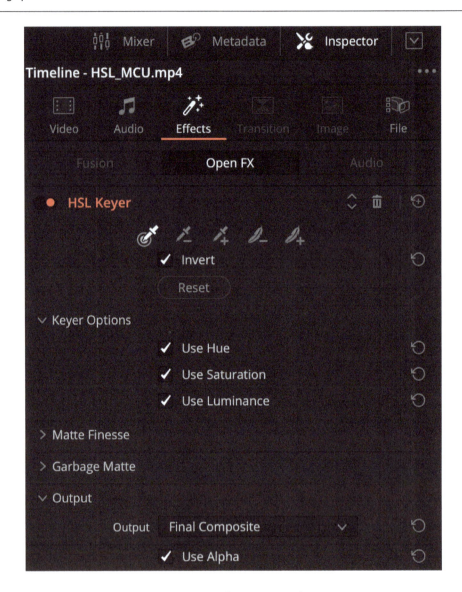

Figure 9.16: HSL Keyer controls

6. We now have our initial key. However, the remaining parts of the green screen have only been partially keyed (*Figure 9.17*), leaving dark patches of green. These patches are dark green as the green that is left is now semi-transparent, showing the black background behind (green mixed with black is dark green):

Figure 9.17: After initial selection (Pick tool)

We can use the **Add Soft** selection tool (feather with a + sign next to it) to feather the edge of the key color that has been selected to encompass the partially keyed parts of the clip (*Figure 9.18*).

Select the **Add Soft** tool and drag on the **Viewer**, starting with the remaining unkeyed parts of the background and finishing at the edge of your subject (that is, drag from the dark green at the top of the frame above their head to the edge of their hair):

Figure 9.18: After Add Soft selection

Now, we have a much better result.

7. Now that you have removed most of the background, you can check your selections by changing your **Output** value to **Alpha Highlight B/W** to see the finished result (*Figure 9.19*). Any area that is pure black is completely removed. Any area that is white is completely opaque, whereas areas that are gray are semitransparent:

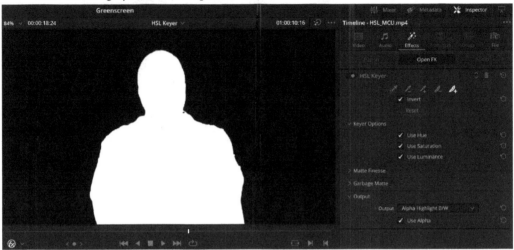

Figure 9.19: Alpha Highlight B/W

8. If you need to remove any gray areas, you can click on the **Matte Finesse** dropdown to reveal the **Matte Finesse** controls, which operate in exactly the same way as the **Matte Finesse** controls in the **3D Keyer.** Use these controls until the image is completely black and white with no gray showing.

9. Don't forget to change your **Output** value to **Final Composite** to see the finished result (*Figure 9.18*).

You may notice that the only green that remains is around the edges of the subject's clothes, face, and hair. This is tricky to select without removing the subject as well. Unlike the **3D Keyer**, the **HSL Keyer** does not have a **Despill** slider to remove this green spill.

Resolve has an extra **Resolve FX** plugin that will help with this task.

Removing chroma key background spill using the Despill Resolve FX Color plugin

The **Resolve FX Color: Despill** plugin is useful for removing despill from clips that have been chroma-keyed using chroma keyers that do not have a built-in despill function. The **Despill** plugin is also useful if you have been given a clip that has already been chroma-keyed but still has colored light spilling onto the subject from the original green screen or blue screen background that has already been removed.

For our clip, we will use the **Despill** plugin to remove the green spill that the **HSL Keyer** could not remove.

Applying the Despill plugin

Just as with the **3D Keyer**, you can apply the **Despill** plugin on either the **Cut** or **Edit** page.

After locating the **Despill** plugin in **Effects** > **Video** > **Resolve FX Color** > **Despill**, you can apply the **Despill** plugin using any of the same methods described for the other keyers.

Adjusting the Despill plugin

Now that you have applied the **Despill** plugin, you can see an instant difference as the plugin has automatically removed the green spill from our subject's clothes and face.

Let us now see what options we have in the **Despill** plugin to help us refine how we remove background spill:

1. On the **Edit** page, open the **Inspector** to reveal the **Keyer**'s controls in the **Effects** panel (**Inspector** > **Effects** > **Despill**).

2. **Despill** has only two main controls—**Key Color** and **Strength**:

    *   **Key Color**: This drop-down menu allows you to select the color of the chroma-key spill that you want to remove. The default setting is green as this is the most common chroma-key screen color. However, the other options of red and blue can be useful if the chroma-key screen is one of these colors.

    *   **Strength**: This slider helps reduce the strength of the despill effect. This is good if you want to reintroduce some of the original color to get a better key. For most situations, this setting can be left at the default setting of 1.000, which is full strength. Any value lower than 1.000 reduces the strength of the color removal:

Figure 9.20: Despill controls

Now that you have removed the green background, all that remains to do is add your own background, as you did for the previous **Timeline**.

Feel free to find your own video or photo backgrounds on websites such as *Pixabay* (`https://pixabay.com/videos/search/background/`) and try them out to find one that suits the look you want.

If you create presentation or training videos, you can export your presentation slides as videos that can be played in the background of your video presentation as you talk to your viewers about your specialist subject. An example use case is a business owner talking to the camera with slides of key products and sales figures behind them.

Summary

You now know how to set up a greenscreen shoot as well as how to remove the background in Resolve using tools such as the **3D** and **HSL Keyer**s. You also know how to apply the **Despill** plugin to remove background spill as well as **Matte Finesse** controls to better refine your key.

This will help you when you want to add different artificial backgrounds in your videos, such as slide presentations to help in presenting a video on a specific topic.

Here is what you have achieved in *Chapter 9*:

- Learned how to shoot videos for a good greenscreen effect
- Used **Resolve FX** keyers to create greenscreen effects on the **Edit** page using the **3D Keyer**
- Used **Resolve FX** keyers to create greenscreen effects on the **Edit** page using the **HSL Keyer**
- Used the **Despill Resolve FX** plugin to remove green spill

In *Chapter 10*, we will look deeper into creating composite images by adding split-screen and picture-in-picture effects.

Questions

1. True or false? The lighting of our greenscreen does not affect our ability to get a good key.
2. True or false? The **3D Keyer** is only for keying 3D images.
3. True or false? The **HSL Keyer** does not have a despill function.
4. True or false? The **Despill** plugin can only be used to remove the spill from green and blue screens.

Further reading

Here are some websites where you can delve deeper into the art of using green screen:

The art of matte painting:

- https://www.rocketstock.com/blog/visual-effects-matte-paintings-composited-film/
- https://www.vanas.ca/blog/digital-matte-painting-and-its-role-in-the-film-industry
- https://www.liveabout.com/matte-painting-in-film-4690782

A free course on the history of matte painting:

- https://www.futurelearn.com/info/courses/vfx-for-filmmakers/0/steps/13286

Copyright-free video backgrounds:

- https://pixabay.com/videos/search/background/

10

Split Screens and Picture-in-Picture

Have you ever wondered how sports broadcasts overlay one video over another so that you can watch a close-up of a footballer's tackle in a small window, while we see the commentators' reactions in the video behind? This effect is also used in news broadcasts and even some movies to show the action happening concurrently. If you are making a wedding video, you can show the groom waiting and show the bride preparing at the same time to show two different sides of the same story. This can be achieved with the help of a split screen.

In this chapter, we will look at creating similar split-screen effects using composite footage and video resizing, as well as **picture-in-picture** effects using the *Resolve FX Transform Video Collage plugin* on the **Cut** page.

In this chapter, we are going to cover the following main topics:

- A simple split-screen effect
- Composite clips
- Picture-in-picture effects – **Create Tile**
- Picture-in-picture effects – **Create Background**

DaVinci Resolve has several different ways we can create split screens by layering videos on the **Cut** page.

We will begin with the simplest and progress to more involved techniques toward the end of the chapter. For each of these techniques, we are going to use the wedding video footage we used in *Chapter 7* and *Chapter 8*.

Technical requirements

To follow along with the exercises in this chapter, download the following project archive, which contains all the clips needed: `https://packt.link/B5bqz`

A simple split-screen effect

Creating a simple split-screen effect is not unique to Resolve and has been used for decades in cinema to show two perspectives of a story at the same time. For example, in early cinema, it was often used to show two characters in different locations having a phone conversation with each other. You can also use split-screen effects to show a close-up of a product at the same time that a presenter talks about it.

The following approach can be used in any editing software and is a good way to understand how layering video works. Let us take the video of our bride and groom and show them cutting the cake with different camera angles, side by side as a split screen, to highlight the event to our audience:

1. Open the *Wedding* project on the **Cut** page.

2. Create a new **Timeline** (*CMD + N*) in the **Media Pool** and call it `Split_Screen`; then, select two video tracks in the **Create New Timeline** pop-up window (*Figure 10.1*).

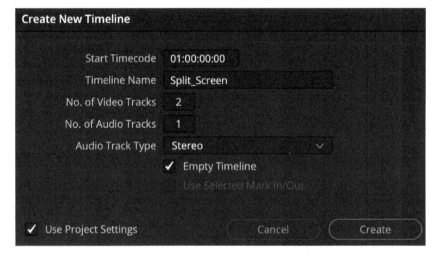

Figure 10.1: The Create New Timeline Window

3. Load the clip of the groom and bride cutting the cake (`MVI_2580.MP4`) from the **Media Pool** into the **Viewer**. Set the in point at `05:05:02:02` and the out point at `05:05:25:17`.

4. Drag the clip of the groom from the **Viewer** onto Track 1 on the **Timeline** (or use the **Append** button in the toolbar).

5. Load the second clip of the groom and bride cutting the cake (A002_04042018_C039. mov) from the **Media Pool** into the **Viewer**. Set the in point at 20:18:28:02. There's no need to set an out point on this one, as we will use the end of the clip as an out point.

6. Drag the second clip of the groom and bride cutting the cake from the **Media Pool** onto Track 2 above Track 1 on the **Timeline** (or use the **Place on Top** button in the toolbar). This will hide the first clip. There's no need to worry about the first clip; the next step will reveal it again.

7. We now need to resize the second clip so that we can see the first clip underneath. Select the clip in Track 2.

8. Click the **Tools** button to reveal the tools under the **Viewer**.

9. Select the **Transform** tool (*Figure 10.2*).

10. Move the top video layer to the left by changing the *x* position slider to have a negative value (such as −207.000, as shown in *Figure 10.2*). This will move the clip to the left, revealing the clip underneath. We now need to crop the right-hand side of the top clip to reveal more of the clip underneath.

Figure 10.2: The Transform tool

11. Select the **Crop** tool (*Figure 10.3*).

12. Move the **Crop Right** slider until it crops the right-hand side of the top image, allowing the image underneath to fill half of the frame (a value of approximately 37.000 should be fine, as shown in *Figure 10.3*). This has now created a split-screen effect, where we see the action happen side by side from two different angles. Remember those 1970s cop shows? They used a very similar effect.

Figure 10.3: The Crop tool

13. Click the **Tools** button again to hide the tools.

14. Play back the **Timeline** to review the effect.

Although we have successfully created a split-screen effect, the two clips are playing back out of sync with each other. This would be fine if they were totally different scenes; however, it is confusing, as both clips were filmed at the same time.

This presents us with an opportunity to show another feature in Resolve called **Auto Align**.

Auto-aligning clips on the Timeline

To be able to sync the timing of two clips that are already on the **Timeline**, we need to pop over to the **Edit** page:

1. On the **Edit** page's **Timeline**, select both clips that you want to sync.

2. Right-click on one of these clips and select **Auto Align Clips** > **Based on Waveform**. You can also find this function in the **Clips** menu – **Clips** > **Auto Align Clips** > **Based on Waveform**. A progress bar will appear to show the progress of Resolve matching up the clips on the **Timeline**, based on the audio waveform (*Figure 10.4*).

Figure 10.4: Auto Sync Audio

That is it – you are now done. You should now see that the clip on Track 2 has now moved along the **Timeline** to match up with the clip on Track 1. If you play them back, you will see that both videos (including their audio) match up perfectly frame by frame.

You can now tidy up this composite by trimming the end of the clip on Track 1 to be the same length as the clip on Track 2. The color between the two clips is different from each other, but we can fix this in the next chapter.

Now that we have synced the clips together, we do not want to separate them by mistake. We may also want to apply effects to both of the clips at once, rather than tweaking each clip individually in the composite. One way to both protect a composite and edit it as a whole is to create a compound clip.

> Key concept – composite
>
> A composite is where a single image is a combination of more than one image. So, the split screen we have just completed is a composite, as well as the greenscreen effects in the previous chapter, as well as the picture-in-picture effects in the rest of this chapter.

Compound clips

You can take a series of clips on the same track on the **Timeline**, or clips stacked on top of each other on different tracks on the **Timeline,** and merge them into a single clip. This single clip that has been created from various video and audio files combined into one is called a **compound clip**. The advantage of working with a compound clip is that you can apply effects and transitions to it as if it were a single clip, without having to apply the same effect several times to its parts. So, anything you can do to a normal clip, you can do to a compound clip.

Creating a compound clip

Creating a compound clip is easy:

1. Select all the audio and video clips on your **Timeline** that you want to turn into a compound clip. You can *Shift + click* on each one to select them or, in our case, use *Command + A* (Mac) or *Control + A* (Windows) to select all the clips on the **Timeline**.

2. Right-click on any of the selected clips and choose **New Compound Clip...** at the top of the pop-up menu.

3. Name your compound clip in the dialog box (*Figure 10.5*).

Figure 10.5: The New Compound Clip dialog box

4. Let us rename ours from `Compound Clip 1` to `Split Screen`.

5. Click the **Create** button (*Figure 10.5*).

We now have a compound clip on the **Timeline** replacing all the clips we selected. There is a new icon at the bottom left of the clip (it looks like a stack of video frames on top of each other) to show that this clip is a compound clip.

Resolve has also automatically created a copy of our compound clip and placed it in the **Media Pool**.

You may be concerned that you can no longer access your individual clips. Don't worry – we can access the individual clips by right-clicking on our compound clip and selecting **Decompose in Place > Using Clips Only** from the pop-up menu. This will return all of the clips to their original state on the **Timeline** before we turned them into a compound clip.

There is a Resolve FX effect called **Video Collage** that can be used to create more complex picture-in-picture effects. **Video Collage** can be found in the **Effects** panel on either the **Cut** or **Edit** page. Let us look at the two different options that **Video Collage** gives us to create these split-screen effects.

Video Collage | Create Background

In the **Video Collage** effect, the **Create Background** option uses a top layer of video as the background layer and punches holes in this layer to show the video on the layers underneath (*Figure 10.6*). These holes appear as video tiles on the final composite video.

Top Video layer with holes cut out

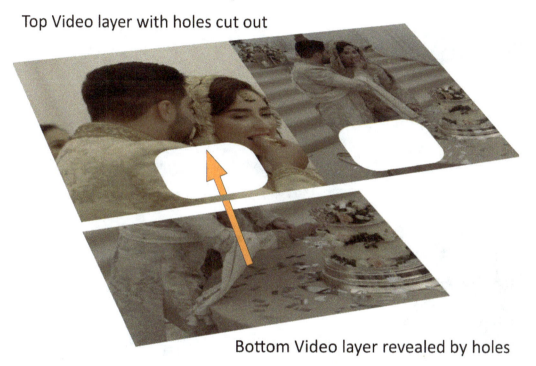

Bottom Video layer revealed by holes

Figure 10.6: A Create Background example

We will be using the split-screen compound clip we created in the last exercise to overlay close-up shots of the cake being eaten.

We are going to use the **Cut** page for the following exercise; however, the effect works in exactly the same way on the **Edit** page.

Preparing our Timeline for Create Background

First, we need to add all the video clips we will use for the picture-in-picture effect:

1. Right-click on the Split_Screen **Timeline** in the **Media Pool**, select **Duplicate Timeline**, and rename it Create_Background. We will use this new **Timeline** for our new effects without losing our original split-screen effects.

2. With the Create_Background **Timeline** open, right-click on the video track header and select **Add Video Track**. You should now have three video tracks.

3. Drag the composite clip from **Video Track 1** to **Video Track 3** (*Figure 10.7*).

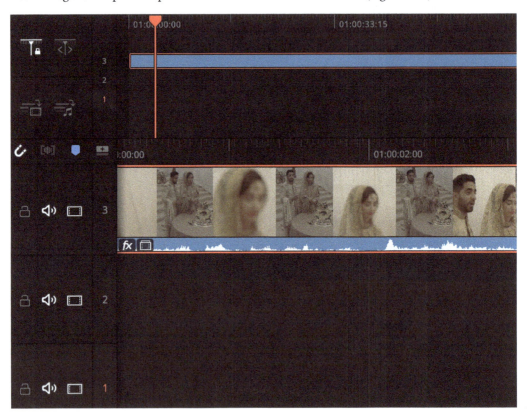

Figure 10.7: The composite clip on Video Track 3

We have now made space to add our new video clips onto the **Timeline**. Let us now add the new clips.

4. In the **Media Pool**, find both clips named MVI_2573.MP4 and MVI_2577.MP4. Drag each one onto a separate empty track on the **Timeline** (*Figure 10.8*).

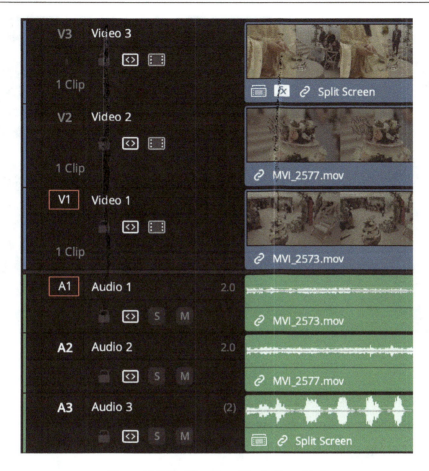

Figure 10.8: The Timeline with all clips added

Now that we have prepared our **Timeline** for **Video Collage | Create Background**, let us apply it to the **Timeline**.

Applying the Video Collage | Create Background effect

Make sure that you have selected the topmost video clip on the **Timeline** before applying the **Video Collage** effect, as you will be punching holes in this video layer to reveal the other clips on the video tracks underneath.

Locate the **Video Collage** effect in **Effects > Video > Resolve FX Transform > Video Collage**. From here, you can apply the **Video Collage** effect (just like we did with the keyer in *Chapter 9*) to the selected clip in the **Timeline** using any of the following methods (*Figure 10.9*):

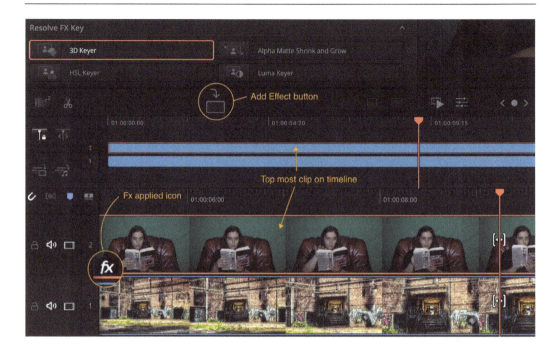

Figure 10.9: Applying effects on the Cut page

- Double-click on **Video Collage** in the **Effects** panel

- Drag and drop the video collage from the **Effects** panel to the selected clip in the **Timeline**

- (On the **Cut** page only) Click the **Add Effect** button (located at the bottom of the **Effects** panel)

Once the video effect has been applied, note that a grid of four tiles appears, showing the video that is on the track directly below.

Let us adjust the number and placement of these tiles to get the effect we want.

Adjusting Video Collage | Create Background

Now that we have applied our video collage, it shows four video thumbnails by default. As we only need two images of the cake, let us change the settings in the **Inspector** window to only show two video thumbnails of the cake from different angles:

1. Open the **Video Collage** control panel in the **Inspector** window (**Inspector** > **Effects** > **Open FX** > **Video Collage**), as shown in *Figure 10.10*.

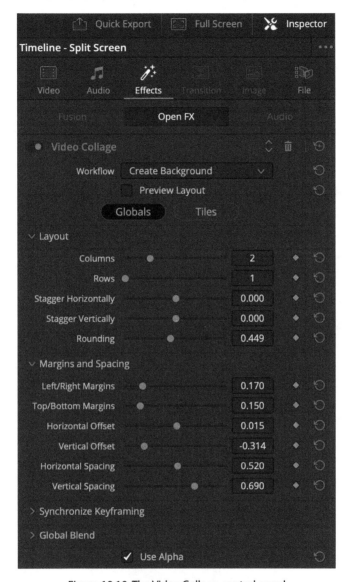

Figure 10.10: The Video Collage control panel

2. With the **Globals** button selected, under **Layout**, move the **Rows** slider to the left to reduce the rows from 2 to 1. You will now have only two tiles rather than four (*Figure 10.10*).

3. Move the **Rounding** slider to the right to make the corners of the tiles rounder and less square (*Figure 10.10*).

4. Select the **Preview Layout** checkbox to make it easier to see the tiles against the background. This turns the tiles into yellow layout placeholders.

5.   Under **Margins and Spacing**, move both the **Horizontal Spacing** and **Vertical Spacing** sliders to make both tiles smaller by the same amount.

6.   Move the **Vertical Offset** slider to the left to move both tiles to the bottom third of the video frame.

7.   Move the **Horizontal Offset** slider to line up the tiles so that they are spaced evenly on either side of the vertical line in our `Split_Screen` clip.

8.   Select the **Tiles** button. This will give us the control to change the tiles individually (*Figure 10.11*).

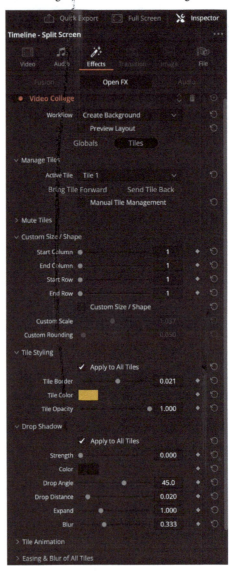

Figure 10.11: The Video Collage Tiles controls

9.   Under **Tile Styling**, move the **Tile Border** slider to the right to create a border around your tile (*Figure 10.11*).

10. Click on the **Tile Color** swatch (*Figure 10.11*) and choose a color from your computer's color picker that suits the video. I have chosen a more muted golden yellow color to go with the color scheme of the bride and groom.

11. Under **Tile Animation** in the **Animate** dropdown, select **Intro & Outro**. Under it, select any combination of the **Fly**, **Shrink**, **Rotate**, and **Fade** checkboxes to choose how to animate the appearance of the tiles at the start and end of the video clip (*Figure 10.12*).

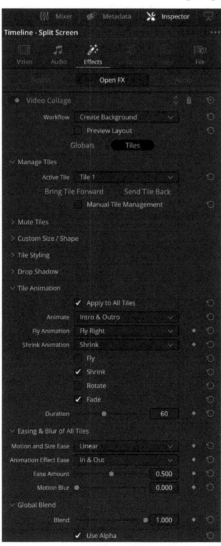

Figure 10.12: The Tile Animation controls

12. Deselect the **Preview Layout** checkbox to reveal how the tiles look on your video and play back the video to test the tile animations. After reviewing your video, make any adjustments that are necessary.

> **Top tip – Globals versus Tiles**
>
> The controls with the **Globals** button selected allow you to change the position and look of all the tiles at once, whereas the **Tiles** button reveals controls that allow you to change the properties of each individual tile.

Now that we have our tiles appearing how we want them to look (*Figure 10.13*), we need to line up the video underneath to fill the tile frames.

Figure 10.13: The tiles in their final position

Repositioning the videos under the tiles

To do this, we will use the same **Transform** tools (*Figure 10.14*) we used to create our first split-screen effect earlier in the chapter:

1. Select the clip on **Video Track 2**, and use the **Transform** and **Crop** tools as described at the start of this chapter to reposition the video on Track 2 to line up with the left-hand video tile.

2. Select the clip on **Video Track 1**, and use the **Transform** and **Crop** tools as described at the start of this chapter to reposition the video on Track 1 to line up with the right-hand video tile.

3. Trim the clip on Video Track 1 to be the same length as Video Track 2, and you're done!

Figure 10.14: Repositioning the video clips using the Transform controls

Now you have a video with two tiles showing two separate videos in miniature at the bottom of your video frame (*Figure 10.15*).

Figure 10.15: The final result

This is just the starting point of how to use **Video Collage | Create Background**. Feel free to explore the other controls in the **Inspector** window to change the look and feel of the tiles to your own taste.

We will now look at another way of using **Video Collage** to create picture-in-picture effects, using the **Create Tile** workflow option.

Video Collage | Create Tile

In the **Video Collage** effect, **Create Tile** uses a layout that you design rather than the pre-made layout design of **Create Background**. **Create Tile** is a bit more work to set up; however, it gives you more options for the design of your layout.

Preparing our Timeline for Create Tile

This time when preparing our **Timeline** for **Create Tile**, we will put our background video on the bottom video track and lay our other clips on tracks above it:

1. Duplicate our **Split_Screen** (**Timeline**) and rename it **Create_Tile**.

2. In the **Timeline**, add two extra video tracks (tracks 2 and 3) above the composite clip that is already on Track 1.

    With **Create Tile**, we leave the background video layer (i.e., our composite clip) on the lowest track (**Video 1**), as shown in *Figure 10.16*.

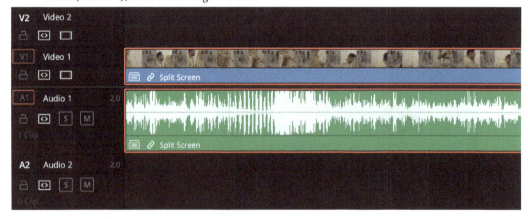

Figure 10.16: The composite clip left on Track 1

3. In the **Media Pool**, find the other clips as before – MVI_2573.MP4 and MVI_2577.MP4. Drag each one onto a separate empty track on the **Timeline**.

Now that we have prepared our **Timeline** for **Video Collage | Create Tile**, let us now apply the effect to the **Timeline**.

Applying the Video Collage | Create Tile effect

Applying the **Create Tile** effect is exactly the same as applying the **Video Collage | Create Background** effect:

1. Make sure that you have selected the topmost video clip on the **Timeline** before applying the **Video Collage** effect.

2. Apply the **Video Collage** effect, just like we did for the **Create Background** effect.

3. Once the **Video Collage** effect has been applied, note that the usual grid of four tiles appears, showing the video that is on the track directly below.

Let us adjust the number and placement of these tiles to get the effect we want.

Adjusting Video Collage | Create Tile

Now that we have applied **Video Collage**, let us change the settings in the **Inspector** window to only show two video thumbnails of the cake from different angles:

1. Open the **Video Collage** control panel in the **Inspector** window (**Inspector** > **Effects** > **Open FX** > **Video Collage**).

2. With the **Globals** button selected, select **Create Tile** in the **Workflow** drop-down menu.

    You will now see that our grid of four tiles has disappeared and been replaced with only one tile, which is a thumbnail version of the current video track, using the video track directly below as the background.

    Let us disable **Video Track 2** so that we can see **Video Track 1** as the background.

3. Select the **Disable Track** button (the *white film strip* icon) on the **Video Track 2** header (*Figure 10.17*). We can now see our bottom track as the background. The **Disable Track** button will now turn red, and the clips on the disabled **Timeline** track will be grayed-out as a visual reminder that the track has been disabled.

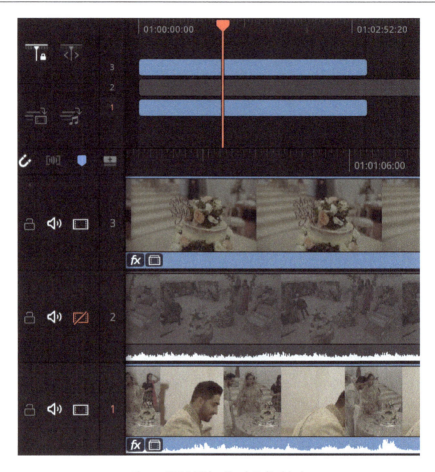

Figure 10.17: Video Track 2 disabled

4.   Select the **Preview Layout** checkbox in the **Video Collage** panel in the **Inspector** window so that we can design our new layout.

5.   Under **Margins and Spacing**, move both the **Horizontal Spacing** and **Vertical Spacing** sliders to make both tiles smaller by the same amount.

6.   Move the **Vertical Offset** slider to the left to move both tiles to the bottom third of the video frame.

7.   Move the **Horizontal Offset** slider to line up the tiles so that they are spaced evenly on either side of the vertical line in our Split_Screen clip.

8.   Select the **Tiles** button. This will give us the control to change the tiles individually.

9.   Under **Tile Styling**, move the **Tile Border** slider to the right to create a border around your tile.

10. Click on the **Tile Color** swatch and choose a color from your computer's color picker that suits the video. As before, I have chosen a more muted golden yellow color to go with the color scheme of the bride and groom.

11. Under **Tile Animation** in the **Animate** dropdown, select **Intro & Outro** and any combination of the **Fly**, **Shrink**, **Rotate**, and **Fade** checkboxes to choose how to animate the appearance of the tiles at the start and end of the video clip.

12. Deselect the **Preview Layout** checkbox to reveal how the tiles look on your video and play back the video to test the tile animations. After reviewing your video, make any adjustments that are necessary.

So far, all of the adjustments we have made to our layout have been the same as we did for the **Create Background** workflow, except for needing to disable **Video Track 2**.

We now need to create a second tile. Instead of adjusting all the settings all over again, we can copy them from the first tile and apply them to our hidden video track to create a new tile.

Assigning video tracks to tiles using Paste Attributes

Now that we have perfected the layout and look for our first tile (on Track 3), we need to copy these settings and apply them to our clip on **Video Track 2**.

To do this, we will use the **Paste Attributes** function on the **Edit** page.

1. On the **Edit** page, select the clip on **Video Track 3** and copy it by pressing *Command + C* (Mac) or *Ctrl + C* (Windows).

2. Re-enable Track 2 using the **Enable Track** button (*Figure 10.18*).

Figure 10.18: Re-enabling Track 2

3. Select the clip on Track 2 and press *Option* + *V* (Mac) or *Alt* + *V* (Windows), or in the **Edit** menu, choose **Edit** > **Paste Attributes**.

Paste Attributes

From   MVI_2577.mov
To   2 Clips ∨

KeyFrames

- ● Maintain Timing   ○ Stretch to Fit

− **Video Attributes**

☐ Composition Mode   ☐ Opacity   ☐ Position
☐ Rotation Angle   ☐ Anchor Point   ☐ Pitch
☐ Yaw

☐ Zoom   ☐ Scale X   ☐ Scale Y

☐ Crop   ☐ Left   ☐ Right
☐ Top   ☐ Bottom   ☐ Softness

☐ Flip   ☐ V Flip   ☐ H Flip

☐ Retime Process   ☐ Motion Estimation

✓ Plugins   ☐ Color Correction ☐ Fusion Effects

☐ **Audio Attributes**

☐ Volume   ☐ Plugins   ☐ Equalizer

☐ **Retime Effects**

Ripple Sequence

Cancel   Apply

Figure 10.19: The Paste Attributes window

4. In the **Paste Attributes** window that appears, select **Plugins** and click **Apply** (*Figure 10.19*).

    The tile effect plugin from Track 3 has now been applied to Track 2. We now need to reposition the tile on Track 2 to its final position to be able to see it.

5. With Track 2 still selected, open the **Effects** panel in the **Inspector** window again.

6. Click on the **Tiles** button to reveal the tile controls.

7. Under the **Manage Tiles** section, select **Tile 2** in the **Active Tile** drop-down menu (*Figure 10.20*).

Figure 10.20: The Active Tile selection

You can now see both tiles, but when you play the video back, the second tile's outro transition is missing. That is because the track is too long.

8. Trim the clip on Track 2 to be the same length as Track 1, and you're done!

Repositioning the videos in the tile frames

With **Create Background**, we shrunk and repositioned video clips to fit into the frames created by holes in the top layer. **Create Tiles** differs in that a whole clip is shrunk down to be the size of the tile. We don't need to do anything else.

However, if we want to adjust what we see in each tile, then we can use the **Transform** tools, just like we did with the clips in **Create Background**.

You now have two ways you can add tiles to a video to create picture-in-picture effects using the **Video Collage** plugin. As with all effects plugins, it is a good idea to play the **Timeline** back to test the effect and then make any necessary adjustments in the **Inspector** window until you are satisfied with the final result.

Now that you have explored the many ways that you can create split-screen effects in Resolve, you will be able to explore how to use split screens in your own videos to enhance your storytelling.

Summary

We now know how to create split-screen and picture-in-picture effects. These effects are useful for how-to videos where we want to show a picture of the presenter and a close-up of their hands at the same time. Other uses for split screens are shown in the *Further reading* section at the end of the chapter.

We also know how to create compound clips, which are useful for applying effects to a whole group of clips at the same time.

Here is what you have achieved in this chapter:

- You learned how to create a split screen by resizing and recomposing composite video footage on the **Cut** page

- You learned what a compound clip is and how to create one

- You learned how to create a picture-in-picture effect using Resolve FX's **Video Collage | Create Background** effects

- You learned how to create a picture-in-picture effect using Resolve FX's **Video Collage | Create Tile** effects

In *Chapter 11*, we will look at how to adjust the color of our clips so that they match and create a visual style for our video.

Questions

1. True or false? You can't create a split-screen effect on the **Cut** page. You will need to use the **Edit** page instead, as it is a more advanced effect.

2. True or false? Once you have created a compound clip, you can no longer access the individual clips.

3. True or false? The Resolve **Video Collage | Create Background** effect is applied to the bottom video layer.

4. True or false? The Resolve **Video Collage | Create Tile** effect is copied from the top track to each clip on the different video tracks underneath using **Paste Attributes**.

Further reading

Here are some extra web links that look at practical uses for split-screen effects:

- `https://www.premiumbeat.com/blog/split-screen-editing-and-composing/`

- `https://nofilmschool.com/2018/03/split-screens-watch`

- `https://www.theguardian.com/film/2022/apr/29/youve-been-reframed-how-playing-with-split-screen-and-aspect-ratio-went-mainstream`

11
Enhancing Color for Mood or Style

One of the most common aspects of filmmaking I have noticed that sets professional-looking videos apart from beginner ones is the use of color to tell a story.

Color is important for filmmaking on so many different levels, and there are whole books written about it.

Rather than replicate what is written in those books, I aim to summarize the color theories so that you can begin to understand how to use color in your videos.

So, in this chapter, we will introduce color theory as used in Hollywood films. Then, we will use **Resolve Color FX** on the **Cut** page to transform, compress, and stabilize color. We will then introduce you to the automatic color tool on the **Cut** page and look at the **Color Management** settings in **Project Settings** to help automatically optimize your image there.

In this chapter, we are going to cover the following main topics:

- Understanding basic color theory
- Stabilizing color
- Contrast pop
- Auto Color
- DaVinci Resolve Color Management (DRCM)

Technical requirements

To follow along with the exercises in this chapter, download the following project archive, which contains all the necessary clips: `https://packt.link/B5bqz`

Let us begin with a basic understanding of color theory.

Understanding basic color theory

Color can be measured and described in many ways, probably more than any other medium. There are whole topics dedicated to color: science, psychology, sociology, and of course, technology. Let me explain each of these in relation to color and why they are important.

Color science

Color reaches our eyes through visible waves of electromagnetic radiation from the Sun and is measured in **nanometers** (**nm**). Our eyes, along with our brains, translate light into color. The perception of each color we see is based on a different wavelength of this radiation, with red (740 nm) being the longest wavelength we can see and violet (380 nm) being the shortest (*Figure 11.1*):

740nm   380nm

Figure 11.1: Wavelengths of visible light (image: Lance Phillips)

Artificial light mimics these wavelengths of the Sun. All the color that we see with our eyes is light either reflected off surfaces or emitted by a light source.

The retina in our eyes is most sensitive to the red, green, and blue wavelengths of light. Our brain processes these wavelengths and mixes them together to create all the other colors we see.

Additive (RGB) color model

The way we see color is called the **Additive Color Model**, as we add different strengths of **Red, Green, and Blue** (**RGB**) light together to create the colors we see. RGB light mixed equally gives us pure white light. This is also the basis of the technology of how our cameras capture color (RGB light sensors) and how our screens portray color using red, green, and blue pixels. Essentially, any technology that emits images using light uses the RGB Additive Color Model. The following diagram shows how the RGB colors mix to create other colors (*Figure 11.2*):

Figure 11.2: The RGB color model (image: Lance Phillips)

Subtractive (CMYK) color model

You may think, well, the color ink in my printer is not RGB, it is cyan, magenta, and yellow. This is the **Subtractive Color Model** and uses **Cyan, Magenta, and Yellow** (**CMY**) to create colors on printed images where the RGB light reflects off the image for us to see it. This Subtractive Color Model is labelled **CMYK** for short where the "K" stands for black ink as it is the key color used in printing.

CMYK is a Subtractive Color Model. So, in a printer using the CMYK model, the ink absorbs (subtracts) some colored light while it reflects the remaining colored light. For example, magenta ink absorbs green light and reflects blue and red light; so blue and red mixed gives us magenta. The following diagram shows how the CMYK colors mix to create other colors (*Figure 11.3*). Essentially, CMYK is a color model used to describe any image that uses ink to create colors. As ink is not a light source, it needs reflected light to create these colors for us to see them:

Figure 11.3: The CMYK color model (image: Lance Phillips)

If you look at *Figure 11.2* and *Figure 11.3*, you will notice that the RGB and CMYK color models combine to create a rough diagram of the full spectrum of visible light.

Color temperature

Not all white light is pure white. As all light is energy transmitted as radiation, the hotter and more powerful the light source, the bluer the light. This is evident when you turn on a gas burner: the flame starts orange and then moves from yellow to blue as you increase the strength of the flame and the temperature. Hence, color temperature is the color a white light gives based on its heat output.

For example, the light a slow-burning campfire emits is orange, an indoor house lamp is yellow, and the Sun on an overcast day is blue. Our brains compensate for this slight color tint and give us the perception that all these sources of light are white.

However, a camera records the color of this white light as it sees it. To compensate for this, most cameras will have a **White Balance** setting that, when enabled, will remove the color tint from the whole image to show white objects in the image as pure white (i.e., if an image is too blue, the camera will compensate by adding orange and vice versa). Additionally, we can use **Resolve** to compensate for this change in color temperature with tools such as **Color Stabilizer**, which we will describe how to use later in this chapter.

Essentially, the color of white light is described as its *color temperature* and is measured in degrees Kelvin (or °K):

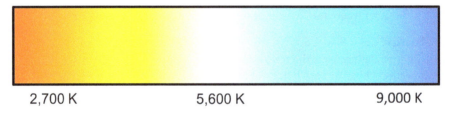

2,700 K   5,600 K   9,000 K

Figure 11.4: Color temperature in degrees Kelvin (image: Lance Phillips)

You may not have noticed, but even modern household bulbs now have their color temperature written on them to help us buy a light that suits what we use it for. A yellow lightbulb will have 2,700 K (degrees Kelvin) and a more neutral whiter light will have 5,600 K (*Figure 11.4*).

All film lights are also measured in degrees Kelvin; some allow you to even adjust a dial to set the color temperature of the light to match any other lights (such as household lights) at the film location.

Color psychology

Color influences our mood and can change how we feel and think. This is because color has symbolic meanings that our brain subconsciously interprets.

Unlike the science of color, where blue is hot, and red is cold, we attach different symbolic meanings to these colors. We perceive blue as a cold color and red as a hot color. That is why we use these colors to label hot and cold water taps.

Out of the most recognizable symbolic meanings of color, the most common is that red represents danger. Therefore, stop signs, brake lights, traffic stop lights, and warning signs are all red. However, red can also represent love and passion, hence why Valentine's cards are often red. Green represents calm, nature, safety, and peace, hence why it is used in green traffic lights to say "go" or in eco packaging to show it is good.

How we use color in our videos can affect the emotional response of our viewers, so it has important consideration when designing the look of our film.

Color sociology

Our ability to perceive different colors is also based on our culture. For example, in China, red represents good luck.

Our ability to perceive different colors can be based on the words we use to describe color. The Himba tribe of northern Namibia has many different words for green, but hardly any words for blue. When they were tested for their ability to see different shades of green and blue, they correctly recognized all the shades of green but could not see the difference between cyan and blue.

This shows that the words we culturally use to differentiate color may influence our perception of color.

Color technology

Increasingly, as technology has improved, so has our ability to show a wider range of colors on our computer screens, TVs, and mobile phones. We now even have a name for the number of colors our screens can show: **Dynamic Range**.

Standard Dynamic Range (SDR) screens can show 16.67 million colors, whereas the latest **High Dynamic Range (HDR)** screens can show 1.07 billion colors. That means that an HDR screen can show almost 100 times more color than an SDR screen.

The range of colors a screen can display is often called a **Color Gamut** (gamut is a Middle English word for range). You might have also heard of the term **Color Space**. Color space and gamut are terms that are often used interchangeably. However, color gamut (which only refers to the range of colors) is actually a subset of color space (which, as well as color, includes luminance and tonal ranges).

Color space

A color space is exactly that: a three-dimensional model (space) showing the number of colors and luminance values that a device can record or display. All color spaces used for videos are based on the RGB color model and show the range of colors and brightness levels associated with that color model.

One popular color space is the Rec.709 color space, which contains 16.67 million colors and a luminance of 100 nits, which is 35% of all the colors that the human eye can see. In comparison, a Rec.2020 color space can contain at least 1.07 billion colors at a minimum of 1,000 nits luminance, which is 75% of the total colors that the human eye can see.

So, the Rec.2020 color space (which is used in HDR televisions) is closer to what the human eye sees than SDR (Rec.709) televisions.

> **Key concept – luminance**
>
> Luminance is the measurable light output of an image. It is measurable and can be used to compare the brightness of one digital image to another.
>
> The term brightness, by comparison, is how we as humans perceive the strength of light, and it cannot be used as a comparison tool as it is not measurable.
>
> Luminance is measured in **candelas per square meter** (**cd/m2**). This is the light emitted by the number of candles in a square meter. So, 100 cd/m^2 is the light of 100 candles in a square meter. This measurement is commonly referred to as nits. So, 100 cd/m^2 is the same as 100 nits.

Now that we've understood the basics of color theory, let us look at the controls in DaVinci Resolve that allow us to change and improve the color qualities of a video.

Stabilizing color (DaVinci Resolve Studio version only)

Sometimes, when editing video footage, we might have single camera shots with unwanted changes in exposure and color, such as when a shot moves from outdoor sunlight into indoor light.

Resolve has a Resolve FX Color plugin called **Color Stabilizer** that corrects this and any other unwanted changes in color and exposure within a single clip. Let us see how it works in practice.

Applying the Color Stabilizer effect

This is how you can apply the **Color Stabilizer** effect:

1. Open your *Color.dra* archive.
2. Create a new **Timeline** and call it **Color**.
3. In the bin named **Color** in the **Media Pool**, drag the clip called `MVI_2572.MP4` onto the **Timeline**.
4. Apply the **Color Stabilizer** FX plugin to the clip on the **Color Timeline**, as we have done in previous chapters. You will find the plugin in **Effects** > **Video** > **Resolve FX Color** > **Color Stabilizer**.

Even though we have applied the **Color Stabilizer** effect, it will do nothing until we tell it which frame of the clip we want to use as our reference for the target exposure and color. Let us play through the clip on the **Timeline** to find a frame that has the desired color and exposure.

You will notice that the camera starts outside on a sunny day, where the sunlight lends a blue tone to the scene. Then, the camera moves into the house, where everything goes dark. After the camera operator adjusts the lens filter to compensate for the indoor light, the shot becomes warmer with a yellow tone. Additionally, you will notice that in the same location, the exposure fluctuates frequently as the Sun changes the lighting conditions.

In our project, as we want to retain the warm look of the indoor wedding scene to match the other clips in the video, move the playhead to the area where the exposure is roughly average for the whole clip (in this example, I chose the frame at 01 : 03 : 23 : 00, as shown in *Figure 11.5*, for its warmer color and even exposure):

Figure 11.5: The frame to be analyzed

We are now ready to use the rest of the **Color Stabilizer** controls.

Adjusting the Color Stabilizer controls

Now that we have chosen our reference frame for color and exposure, let us look at the **Color Stabilizer** controls and how to use them to correct the changing color and exposure of light in our clip (*Figure 11.6*):

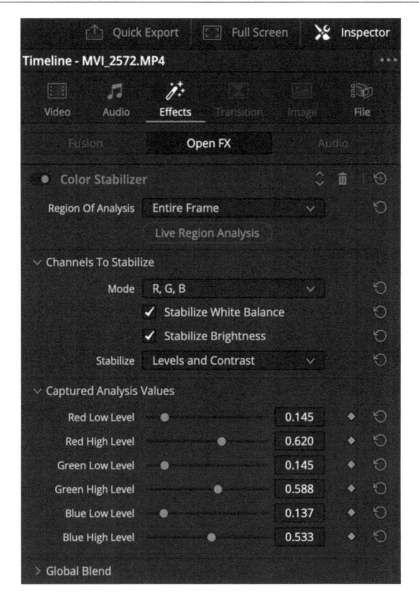

Figure 11.6: The Color Stabilizer controls

1. Select the clip in the **Timeline** and then open the **Color Stabilizer** controls in the **Inspector** panel: **Inspector** > **Effects** > **Open FX** > **Color Stabilizer**.

2. Select **Entire Frame** in the **Region Of Analysis** drop-down menu. This will analyze the entire frame for color and exposure. On the other hand, **Selected Area**, another option in the drop-down menu, will only analyze part of a frame, which is a more advanced method that is beyond the scope of this book.

The whole clip will have its exposure and color adjusted to match the currently selected frame where you have placed your playhead. We need to play through the clip to see the changes.

You can fine-tune the changes by selecting which color channels to stabilize.

3. In the **Mode** drop-down menu, you can select **Balance, Brightness** to balance just the exposure, or select **R, G, B** to balance the color. For this clip, selecting **R, G, B** (remember the RGB color model that we discussed earlier in the chapter) has the best results as it evens out the different color temperatures and tones throughout the shot.

   The **Stabilize White Balance** and **Stabilize Brightness** checkboxes allow you to select or deselect these effects if you have chosen the **Balance, Brightness** mode.

4. Finally, in the **Stabilize** drop-down menu, we can select between **Levels and Contrast**, **Offset**, and **Gain**. This allows you to select how the levels are analyzed and whether Resolve looks to match exposure levels and contrast, offset the whole image to be consistent, or adjust the gain to a consistent level.

   Which option you choose depends upon the image. I tried all three and found that **Gain** gave the most pleasing and consistent results for this clip.

5. Play back the clip to review the results:

Figure 11.7: Outdoor scene before the Color Stabilizer

We can see the results of our **Color Stabilizer** by comparing the footage of the **Color Stabilizer** switched off (*Figure 11.7*) to what the clip looks like with the **Color Stabilizer** switched on (*Figure 11.8*):

Figure 11.8: Outdoor scene after the Color Stabilizer

Now that we have made the exposure and color of the clip more consistent, you will notice that the areas that were previously dark have been raised in exposure, revealing a video of people entering the building, which we could not see as well before since the image was too dark. Also, the resulting image throughout the clip is warmer and has a more consistent color temperature.

However, even though the image is now consistently exposed, it lacks contrast. Thankfully, there is another **Resolve FX Color** plugin we can use, called **Contrast Pop**, to help fix this.

Contrast Pop (Studio version only)

Contrast Pop is a **Resolve FX Color** plugin that allows you to add more (or less) contrast to a selected part of the video clip's tonal range (i.e., you can make the mid-tones have more contrast while leaving the shadows and highlights of an image alone).

Applying the Contrast Pop effect

Apply the **Contrast Pop** FX plugin to the same clip on the **Timeline**, just as we did with the **Color Stabilizer** effect. You will find the plugin in **Effects** > **Video** > **Resolve FX Color** > **Contrast Pop**.

Even though we have applied the **Contrast Pop** effect, it will do nothing until we adjust its settings.

Adjusting the Contrast Pop effect

Let us look at the **Contrast Pop** effect's controls in the **Inspector** panel and how we can use it to improve an image's contrast:

1. Select the clip in the **Timeline** and then open the **Contrast Pop** controls in the **Inspector** panel: **Inspector** > **Effects** > **Open FX** > **Contrast Pop** (*Figure 11.9*):

Figure 11.9: The Contrast Pop settings

2. By default, the **Detail Amount** slider is set to zero, which means no effect has been applied. Moving the slider to the left-hand side (negative number) decreases contrast, creating a softer look, and moving the slider to the right-hand side (positive number) increases contrast, adding more sharpness to the image.

    For this image, we need to move the slider to the right-hand side to add more contrast. We only need to move the slider a small amount (e.g., 0.150) to make a noticeable difference.

3. By default, the **Detail Size** slider is set to 0.500. **Detail Size** lets you select which parts of the image will have their contrast changed, with smaller details to the left-hand side and larger details to the right-hand side. For this image, leaving the **Detail Size** setting at the default value of 0.500 is fine. However, you can experiment with the slider yourself based on your personal preference.

4. The **Low Threshold** slider selects the lower tonal range that the contrast effect is applied from. Moving it to the left-hand side includes more of the shadows in the contrast adjustment, and moving it to the right-hand side excludes more of the shadows from the same adjustment. Adjust this based on personal taste.

5. The **High Threshold** slider selects the upper tonal range that the contrast effect is applied below. Moving it to the left excludes more of the highlights from the contrast adjustment, and moving it to the right includes more of the highlights in the same adjustment. Again, adjust this based on personal taste.

6. The **Softness** slider allows you to soften the transition between the areas of the image that are unaffected and affected by the contrast adjustment. This is good to use if the image starts to have unwanted digital artifacts or noise. In this case, the default softness value of 0.500 is good.

Key concept – digital artifacts

A digital artifact is the unwanted changes to an image after it has been processed by a camera or software such as applying an effects plugin.

An example of one of the most common digital artifacts is noise. Noise appears in an image as unwanted multicolored pixels (usually, stray red, green, or blue pixels) when the gain (brightness) of an image is increased beyond what the image can cope with.

Using **Contrast Pop**, we have now added more contrast to the mid-tones of our image, which focuses our attention on our subjects (which are usually people), separating them from the softer background:

Figure 11.10: Before the contrast pop

We can see the results of our **Contrast Pop** by comparing the footage of the **Contrast Pop** switched off (*Figure 11.10*) to what the clip looks like with the **Contrast Pop** switched on (*Figure 11.11*):

Figure 11.11: After the Contrast Pop is applied

Although the contrast has improved dramatically (*Figure 11.11*) compared to what it was like before (*Figure 11.10*), the image still lacks color. Let us look at the **Auto Color** tool on the **Cut** page to fix the color.

Auto Color

All high-end films on streaming platforms and in the cinema will involve an element of color correction (commonly referred to as *Grading*). A professional colorist (the person who color grades a film) will remove any color cast or tint in the areas of the image that are meant to be pure black or white but have been created by the camera or lights set to the wrong color temperature for the scene (e.g., a scene was shot with reddish white light but the scene is meant to represent the bluish white light of midday Sun). This is an aspect of *Grading* called *Balancing*, as we are making sure that the whites (and blacks) are balanced between having a red or blue color temperature, resulting in a neutral white (pure white with no color tint).

However, you do not need to be a professional colorist to be able to improve your social media videos in Resolve. Resolve has an **Auto Color** tool in the **Tools** section of the **Cut** page that will quickly correct each video clip to correct the color and contrast of the image.

Auto Color uses Resolve's **Neural Engine** to analyze the color in the frame underneath your playhead. Resolve looks for any color imbalance in the blacks and whites of the image and corrects them to be neutral. Resolve then applies this correction to the rest of the clip using the **Neural Engine**.

Let us apply **Auto Color** to our wedding clip:

1. On the **Cut** page, move the playhead to the same frame we analyzed for the **Color Stabilizer** (for me, it's the frame at 01:03:23:00) so that we get consistent results throughout the clip.

2. Click on the **Tools** button (second from the left under the **Viewer** option) to reveal the **Cut** page tools (*Figure 11.12*):

Figure 11.12: Auto Color applied

3. Select the **Color** icon (it looks like a color wheel), which is second from the right underneath the **Viewer** option (*Figure 11.12*).

4. Click on the **Auto Color** button (*Figure 11.12*).

The whole clip has now been corrected, and there is a noticeable improvement in both the color and contrast.

Auto Color is particularly well suited for color-correcting videos for social media as it is optimized for the Rec.709 color space, which is the range of colors most consumer cameras and phones capture their video in.

So far, we have limited ourselves to only the color correction tools that we can find on the **Cut** page. However, there is one final tool, outside the **Cut** page, that we can quickly use to help us enhance the colors in our video. This tool is called **DaVinci Resolve Color Management (DRCM)**. Let's take a look at it next.

Exploring DRCM

Now, I know plenty of professional colorists who find this topic too scary to mention. However, *Blackmagic Design* has made color management in DaVinci Resolve easy to use for all levels of experience.

So, what is **Color Management**? Well, firstly, it is much more than a tool; it is a different way of working. It is where we match the color space the camera recorded to our computer monitor's color gamut and then to the color space of the final video. So far, we have been correcting the color of our video based on what we see in the **Viewer** with no regard to the color the camera captured. This is not always the most accurate approach, as our video could look different on various computer screens and may not look like what the camera saw.

Earlier, we mentioned the Rec.709 color space, which is the range of colors a consumer camera can record. Rec.709 (our input color space from our camera) is also the same color range as an SDR monitor using the standard sRGB color gamut. As social media videos (e.g., YouTube) are currently optimized for display on SDR devices such as phones and computer screens, Rec.709 is also the ideal color space for our final video (the output color space for our exported video).

However, even though this pathway of color from the camera to our computer screen to final output on social media (which could be any form of a screen such as phones, tablets, computers, and more) seems roughly the same, there can be subtle differences in how the color is processed at each stage.

This is where DRCM comes in. It reads the color space of the camera from the imported video clips and changes the color of our video clip on the **Timeline**. DRCM does this to better show on our computer monitor what the original camera saw by adjusting for the differences between the two. If we do not use color management, then our display might not be showing us the most accurate version of what the camera recorded.

Also, color management is good for when we have images recorded from more than one camera (all cameras record color differently). DRCM will recognize the color space that each camera used to record the video clip and adjust the color of each clip on the **Timeline** to display the color on the computer monitor correctly.

Now that we know why using DRCM is useful, let us see how we can enable it:

1. Click on the **Project Settings** icon (the small cog in the lower-right corner of the screen; *Figure 11.13*) to open it:

Figure 11.13: The Project Settings icon

2. Select **Color Management** > **Color Space & Transforms** (*Figure 11.14*):

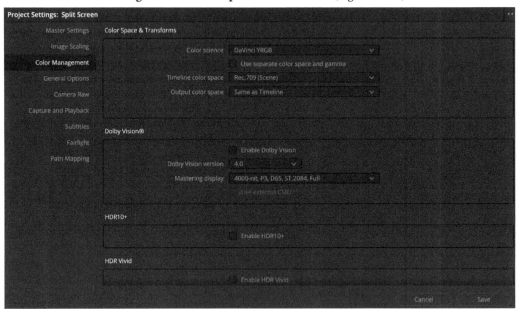

Figure 11.14: The Color Management settings

3. In the **Color science** drop-down menu, select DaVinci YRGB Color Managed (*Figure 11.15*). Leave the **Automatic color management** checkbox selected and the **Color processing mode** option set to SDR. Leave **Output color space** as SDR Rec.709:

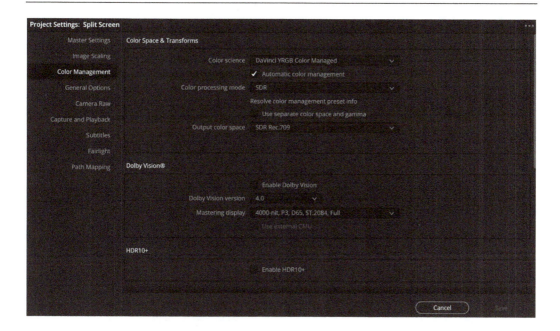

Figure 11.15: After selecting DaVinci YRGB Color Managed

4.   Click on the **Save** button.

You will notice that the clip has now changed to be ever more colorful and vibrant to reflect the original color that the camera captured (*Figure 11.16*):

Figure 11.16: After Color Management has been enabled

DRCM can be enabled at any time; however, it will change how your previous color corrections will look, so it is best to enable this feature before you start to make any other color changes. For example, the Resolve FX **Contrast Pop** plugin might now need to be adjusted (or turned off) so that the image does not have too much contrast.

The color-correcting plugins we have covered in this chapter are just a selection of the basic tools that are available to you in DaVinci Resolve to enhance the color of your videos.

Summary

You now have an overview of basic color theory, which can act as a good foundation for further learning and understanding how to use color in videos. You have used the **Resolve FX Color** plugins, **Color Stabilizer** and **Contrast Pop**, on the **Cut** page to correct the changing color and exposure of the video and enhance the contrast of the clip. You have used the **Auto Color** tool on the **Cut** page to quickly enhance the video's color and, finally, explored DRCM to automatically match the camera's color space. We've done all of this with the overall objective to enhance the color and look of your video.

Here is what you have achieved in *Chapter 11*:

- Understood basic color theory

- Learned how to use Color FX **Color Stabilizer** to change the color of the footage

- Learned how to use Color FX **Contrast Pop** to change the contrast of the footage

- Learned how to use the **Auto Color** tool on the **Cut** page

- Learned how to enable and use DRCM

In *Chapter 12*, we will look at other AI Neural Engine tools, which are only available in the Studio version of DaVinci Resolve, and how they can be used to speed up your edit process.

Questions

1. True or false? The human eye has cyan, magenta, and yellow light-receiving cones in the retina.

2. True or false? You can use the **Color Stabilizer** FX plugin to compensate for changing color and exposure in the same shot.

3. True or false? The **Contrast Pop** FX plugin can only change the contrast of the whole image; it is not possible to change the contrast of only parts of the image.

4. True or false? The **Auto Color** tool is not available on the **Cut** page; you need to go to the **Color** page to use it.

5. True or false? DRCM is found in **Project Settings**.

Further reading

Here are some extra web resources you can refer to regarding color theory:

Color science:

- https://www.amnh.org/explore/ology/physics/see-the-light2/the-color-of-light
- https://www.amnh.org/explore/ology/physics/play-with-color-and-light

Color temperature:

- https://www.studiobinder.com/blog/what-is-color-temperature-definition/
- https://www.lighting.philips.co.uk/consumer/led-lights/warm-led-light

Color spaces:

- https://blog.frame.io/2020/02/03/color-spaces-101/

12

Studio-Only Techniques

In the previous chapter, we started to look at DaVinci Resolve Studio-only tools that use the Neural Engine, such as the **Color Stabilizer** or **Contrast Pop**. As the Neural Engine needs a more expensive computer with a newer and faster processor, the tools that require this are usually only available in the full paid-for Studio version of Resolve and not available in the free version.

In this chapter, we will look at the **Neural Engine's Artificial Intelligence** (**AI**) tools, only available in the Studio version of DaVinci Resolve, that can speed up your edit process. We will look at **Smart Reframe** to recompose your video for different social media formats, **Smart Bins** to quickly organize video footage of people into their own individual bins, and finally, **Face Detection** and how to use it to search for and identify people in your footage.

In this chapter, we are going to cover the following main topics:

- **Smart Reframe** on the **Cut** page
- **Smart Bins** for people on the **Edit** page
- Using **Face Detection** on the **Cut** page

If you want to stay with the free version, then you can skip this chapter, or you can read it to see whether some of these extra features make it worth your while upgrading from the free version to the Studio version of Resolve. The choice is yours!

Technical requirements

You can download the `Lance Jaggers Audition` archive file to work through the exercises in this chapter here: `https://packt.link/B5bqz`

Smart Reframe on the Cut page

Occasionally, when you are shooting your footage quickly, you don't always get the opportunity to frame your shot how you want. One of the advantages of shooting in 6K or even 4K is that you can zoom in on your footage to reframe your shot closer for a 1080p film with no loss of quality. However, this is usually a manually intensive job where, on the **Cut** page, you would use the **Transform** controls under the **Tools** palette to reframe the clip.

If you need a quick turnaround for your video and you have the Studio version of Resolve, you can use the **Smart Reframe** tool, which uses Resolve's **Neural Engine** to recognize the subject of your video and reframe the video to create a shot centered around your subject.

This is particularly useful if you change the aspect ratio of your video from the horizontal 16:9 to the vertical 9:16 or square 1:1 aspect ratios often needed to display videos on social media platforms, such as *Facebook* or *Instagram*.

Let us look at how we can use **Smart Reframe** to resize our video for the social media aspect ratio of 1:1.

We are going to reframe a new project called `Lance Jaggers Audition`. This was shot on an iPhone as a 3840 x 2160 aspect ratio video.

I need to resize it to the square 1080 x 1080 aspect ratio so that I can upload it to *Instagram*. We can use the **Timeline Settings** in both the free and Studio versions of Resolve to resize the aspect ratio of our video.

Changing our Timeline aspect ratio

Let's begin:

1. In the **Media Pool**, duplicate the `Lance Jaggers Audition` **Timeline** (right-click and select **Duplicate Timeline**).

2. Rename the new `Lance Jaggers Audition copy` **Timeline** `Lance Jaggers Audition reframed`.

3. Right-click the `Lance Jaggers Audition reframed` **Timeline** in the **Media Pool** and select **Timelines** > **Timeline Settings....**

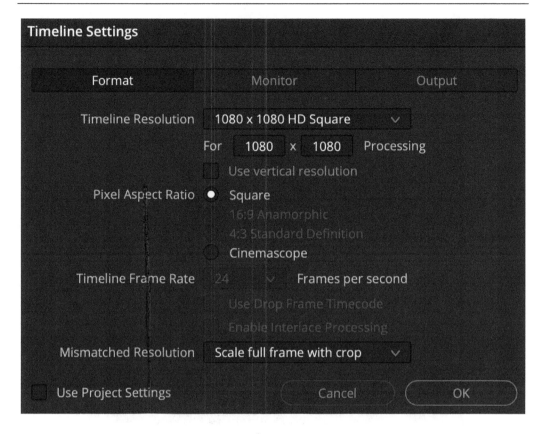

Figure 12.1: Timeline Settings

4. In the **Timeline Settings** window, deselect **Use Project Settings** (*Figure 12.1*); this will enable you to change the aspect ratio of your **Timeline** and, hence, the final video.

5. In the **Timeline Resolution** drop-down menu, choose **1080 x 1080 HD Square** (*Figure 12.1*).

6. In the **Mismatched Resolution** drop-down menu, select **Scale full frame with crop** (*Figure 12.1*).

7. Click the **OK** button to change the resolution of the entire **Timeline**.

We have now changed our video so that the aspect ratio is no longer 3840 x 2160 but, instead, a more suitable square aspect ratio for *Instagram* of 1080 x 1080.

Figure 12.2: The video with a new square aspect ratio but with a cropped subject

Note that the **Timeline** has cropped out the sides of the original video. However, although most of the subjects remain centered within the frame, there are some clips (*Figure 12.2*) in which the subject needs to be more centered in the frame; this is where **Smart Reframe** comes in.

Using Smart Reframe

Smart Reframe uses Resolve's Neural Engine to analyze the video clip for a subject and then place them in the center of the frame throughout the whole of the video clip.

Smart Reframe is not a special plugin or tool but, instead, a setting in the **Transform** panel of the clip's Inspector.

Let us see how we can use **Smart Reframe** to reframe our video content for social media:

1. Select the clip or clips you want to reframe on the **Timeline**.

2. Open the **Transform** settings in the **Inspector** window (**Inspector** > **Video** > **Transform** > **Smart Reframe**).

3. Click on the **Smart Reframe** dropdown to reveal the controls (*Figure 12.3*).

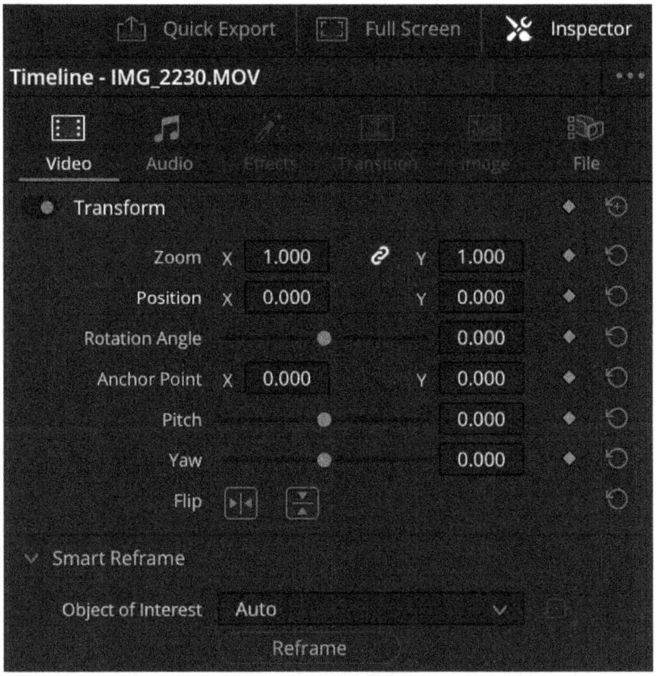

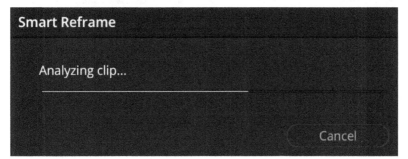

Figure 12.3: The Smart Reframe settings

4. The **Object of Interest** drop-down menu is set to **Auto** by default (*Figure 12.3*).

5. Click the **Reframe** button (*Figure 12.3*). Resolve will now analyze the clip(s) (*Figure 12.4*) you have selected and change the **Transform** settings automatically to reposition your subject within the frame. A popup will appear to show you the progress of the analysis (*Figure 12.4*).

Smart Reframe

Analyzing clip...

Cancel

Figure 12.4: The Smart Reframe analyzing progress bar

You will now have a video where Resolve has analyzed the footage for any movement of the subject and automatically panned the video, keeping the subject within the frame of the new aspect ratio.

When you play back the video, it feels like a camera operator is panning the shot to match the subject's movement while keeping the subject in the center of the frame, as can be seen when comparing the clip before **Smart Reframe** (*Figure 12.2*) to after it being applied (*Figure 12.5*).

Figure 12.5: Smart Reframe applied

If you are not happy with how Resolve has automatically reframed the subject (e.g., Resolve chose the wrong subject to re-frame around), you can manually select a **reference point** for your subject on a clip-by-clip basis.

Manually selecting a reference point

Let us explore how we can select our own subject's reference point for Resolve to reframe the aspect ratio around.

Smart Reframe has automatically placed the subject into the center of the frame. However, we want to give our subject some more looking room and so want to place our subject more to the right of the frame as they are looking toward the left of the frame.

> Key concept – looking/lead room
>
> Sometimes, when framing a subject in a video, we do not always want them to be in the center of the frame.
>
> One instance of this is when our subject is looking into the frame. It is more aesthetically pleasing to have more empty space in front of the subject than behind them; this extra space in front of the subject is called *looking* or *lead* room, as it leads the audience's eyes in the direction that the subject is looking.
>
> Looking room can also be applied to vehicles, not just people, to give the vehicle space to be able to travel across the frame.

To do this, we will select a reference point at the front of the subject's waistcoat, which will result in moving the object slightly to the left as Resolve moves the waistcoat to the center of the frame. Let us explore how to do this:

1. Just like you did with the auto **Smart Reframe** tool, select the clip you want to reframe on the **Timeline** and reveal the **Smart Reframe** controls in the **Inspector**.

2. Under the **Object of Interest** drop-down menu, select **Reference Point**. This will reveal a white target icon to the right of the **Object of Interest** drop-down menu. A reference point bounding box will also appear on the **Viewer**.

   If the reference point bounding box does not show automatically, you can turn it on and off by clicking on the target icon.

3. Reposition this reference point bounding box to your new subject in the video by clicking within the box and dragging it to its new position. Let us do that to move the box so that the center covers the front of the subject's waistcoat against the white shirt sleeve (*Figure 12.6*), as this is a good high-contrast part of the image that Resolve will find easy to reframe.

4. You can also resize the bounding box by clicking and dragging on any of the corners of the box. Let us do that by resizing the bounding box to cover the waistcoat's closed buttons and part of the subject's arm (*Figure 12.6*).

Figure 12.6: Repositioned bounding box

5.   Click the **Reframe** button for Resolve to reframe the aspect ratio around your selected object of interest.

6.   Click on the white target icon (*Figure 12.3*) to hide the reference point bounding box in the **Viewer**.

By placing the waistcoat in the middle of the frame, Resolve has now moved the subject more to the right (*Figure 12.7*).

Figure 12.7: The Smart Reframe result using a new object of interest

This new framing allows the subject to be framed asymmetrically, which can be more visually pleasing to the viewer.

Top tip – smart reframing reference point selection

When choosing a reference point to place the bounding box over, you will get better results if you choose a place where there is a well-defined edge or area of contrast. This enables Resolve to easily recognize the subject throughout the video clip.

In order to place the bounding box over a subject outside of the frame, you can use the **Transform** controls in the **Inspector** window to reposition the clip so that you can see your subject.

We have used **Smart Reframe** to be able to quickly reposition a new aspect ratio around our subject automatically. Let us now look at another Neural Engine tool that allows us to recognize people and put them automatically into their own bins, making it much easier to quickly find the footage we are looking for.

Smart bins for people

So far in our The Wedding video project, we have manually played through each video and looked for clips that feature the bride and the groom and then manually separated them into separate bins. However, we can only have one bin this way, as the bride and groom both feature in the same clips, so we can't separate them into their own bins without creating copies of each clip.

There is a better way to separate clips of different people into their own bins – with the help of smart bins.

Let us enable **Smart Bins** for our The Wedding video project.

Enabling Smart Bins

To be able to use **Smart Bins**, we first need to enable them in the **Preferences** menu:

1. Go to the **Menu** bar and navigate to **DaVinci Resolve** > **Preferences** > **User** > **Editing** > **Automatic Smart Bins** (*Figure 12.8*).

2. Enable the checkbox next to **Automatic smart bins for people metadata** (*Figure 12.8*).

3. Click the **Save** button.

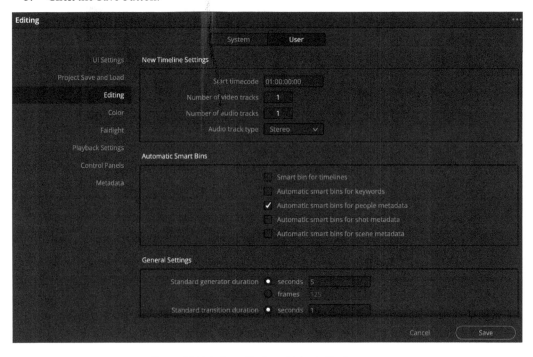

Figure 12.8: Automatic Smart bins in Preferences

On the **Edit** page, a smart bin called **People** has now appeared at the bottom left of our **Media Pool** (*Figure 12.9*).

Figure 12.9: The People smart bin

Note that our **People** smart bin is empty. We now need to search for people in our footage and add their names to each clip. Let us see how we can do this manually by adding metadata.

Adding metadata manually

Let us see where to add the **People** metadata for any clip:

1.   Select any clip in the **Media Pool**.

2.   On the **Edit** page, open the **Metadata** panel (next to the **Inspector** window, as shown in *Figure 12.10*).

3.   Click on the **Sort** icon (the downward-facing arrow) to reveal a drop-down menu of the different metadata groupings.

4.   Select **Shot & Scene**. The **People** textbox is the fourth one down (*Figure 12.10*).

Note that the **People** textbox is empty. If you manually type in a name here, it will create a smart bin with the name of the person you have added in the **People** textbox.

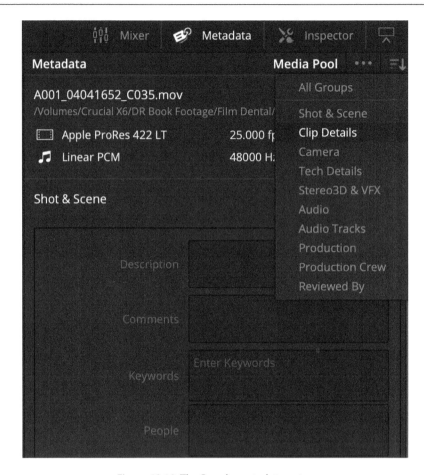

Figure 12.10: The People metadata entry

Now, when you look in the **People** smart bin, you will see that it contains any clips where you have added names to the **People** metadata. Each name you add to a clip will add a sub bin to the **People** smart bin with that person's name on it.

If you select several clips at once in the **Media Pool** and then add names to the **People** text entry box in the **Metadata** panel, then multiple clips can be named simultaneously, saving you time, and all those clips will appear in the smart bin.

However, if you have the Studio version of Resolve, you can save even more time by letting the Neural Engine identify people's faces in each clip and group the clips into different bins for you.

Using Face Detection on the Cut page

Thankfully, with the Studio version of Resolve, we don't need to add the names manually to each clip's metadata. We can use face detection to automatically recognize faces using Resolve's Neural Engine and add their names to each clip as metadata. Unlike typical metadata entries, we can do this on the **Cut** page. For this exercise, we are going to use our The Wedding project and select all the clips in the **Cake** bin:

1. Right-click on one or several clips you have selected in the **Cut** page's **Media Pool**. If you select a bin (with clips in it) as well as a bunch of other clips next to the bin, Resolve will search the clips in this bin as well as the other clips you have selected that are not in the bin.

2. In the drop-down menu, select **Analyze Clip for People**. A progress bar will appear as Resolve analyzes your clips for people's faces (*Figure 12.11*).

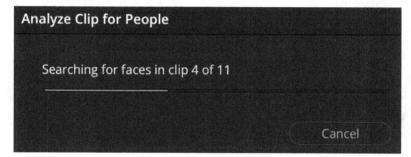

Figure 12.11: The Searching for faces progress bar

After Resolve has searched for and recognized faces in your clip, it groups the faces (*Figure 12.12*) and places them into their own individual bins.

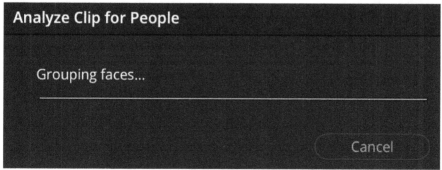

Figure 12.12: The Grouping faces… progress bar

3. Once the analysis is complete, a **People** management window appears (*Figure 12.13*):

Figure 12.13: The People management window

- If an **Improve People Grouping Result** button appears, click this to help Resolve's AI identify faces that it has difficulty with. An **Improve Face Clustering** window (*Figure 12.14*) will appear, asking you to compare two images and identify whether they are the same person or not. Click the **Yes** or **No** button, depending on whether the images are of the same person or not.

Figure 12.14: The Improve Face Clustering window

- The **People** management window (*Figure 12.13*) is split into two sections:

  - All the people Resolve has found and put into groups are listed on the left. Clicking on the **People** group will reveal thumbnails of all the people Resolve has recognized (this is the default view). The other groups listed are for each individual person (e.g., **Person 1**, **Person 2**, etc.) to filter only the thumbnails of clips they feature in. Finally, the **Other People** group shows the people that Resolve could not place into a distinct group.

  - The main body of the window to the right has thumbnails of all the people Resolve has identified, filtered by whatever group you have selected in the list on the left.

4. Select each thumbnail and click on the **Enter name here...** text to rename each of the people that have been identified. As an example, identify the bride and groom and rename the clips as such. The name of the bin on the list will change to reflect the name changes you have made (e.g., **Person 2** has been renamed **Groom** in *Figure 12.15*).

Figure 12.15: Renaming thumbnails

5. Open the **Other People** group by double-clicking on either the **Other People** thumbnail on the right or the **Other People** group listed on the left (*Figure 12.16*).

6. Scroll down until you see a clip where the groom has not been put into a group (*Figure 12.16*). You will see an orange box around a face, where Resolve has identified a face but does not know which person it belongs to.

Figure 12.16: The Other People group

7. Right-click on the thumbnail of the groom and select **Tag As** > **Groom**. This will remove this thumbnail from the **Other People** group and allocate it to the **Groom** group.

8. Once you have finished naming and allocating people to the different groups, you can click the **Close** button at the bottom right of the window.

9. Once you have closed the **People** management window, you may wonder, *how do I get it back again so that I can rename people if I need to?* You can easily reopen the window by choosing **Workspace** > **People** in Resolve's menu bar.

Top tip – correcting the face identification if the results are wrong

If Resolve has wrongly identified a person, you can right-click the thumbnail and select the person from the list if it is an existing person, or select a new person to create a new group for that person.

If there are too many wrong guesses and you want Resolve to start again, you can select **Reset Face Database** in the options menu (the three dots at the top right of the **People** management window).

Unfortunately, at the time of writing, there is no way to see **Smart Bins** on the **Cut** page. So, to see the results of our face recognition, we need to go to the **Edit** page.

In the **Smart Bins** section of the **Media Pool**, note that the **People** smart bin now has a drop-down arrow next to it. Clicking on this arrow will reveal all the **People** smart bins listed by name (including a smart bin for our **Groom** that we have just named). Clicking on any of these smart bins will reveal only the clips that contain that person.

Figure 12.17: People smart bins

Also, if you look at the **Metadata** panel for each clip, you will see that the **People** textbox is now populated with all the names (as keywords) of the people that Resolve's facial recognition identified for that clip (*Figure 12.18*).

Figure 12.18: The Metadata panel with people keywords

The great thing about smart bins is that if we add any **People** metadata to a clip (either manually or through facial recognition), the smart bin will be automatically updated with the new clip.

So, now you will be able to quickly organize your clips in Resolve using **Face Detection** to automatically sort clips of people into their own individual smart bins. You can also quickly change the framing of your clips to suit different aspect ratios for the various social media video platforms.

This chapter was a brief overview of some of the Neural Engine tools available in the Studio version of DaVinci Resolve that will help you create videos for social media and the web more quickly and efficiently. These tools, as well as the others covered in the rest of this book, will give you a good foundation for further exploration in using Resolve for your video editing needs.

Summary

Here is what you have achieved in *Chapter 12*:

- You learned how to use **Smart Reframe** on the **Cut** page to change the video framing for different social media formats
- You learned what a smart bin is and how to use it to create bins automatically based on metadata
- You learned what **Face Detection** does and how to use it to search for and identify people in your footage

Congratulations! You have now completed the book, and you are now better equipped to continue your journey of using Resolve to edit and polish some amazing videos for your business, social media, or the web.

Good luck! I can't wait to see what you create!

Questions

1. True or false? **Smart Reframe** is a Resolve FX plugin found in the **Effects** panel.
2. True or false? Smart bins automatically update their content to show new clips added to the **Media Pool**.
3. True or false? **Face Detection** can recognize faces in your clips and put them into separate bins, but you still need to name each bin with the actual person's name.

Further reading

Here are some articles on video composition to help you when reframing your shots:

- `https://motionarray.com/learn/filmmaking/shot-composition-framing-rules/`
- `https://en.wikipedia.org/wiki/Lead_room`
- `https://neiloseman.com/lead-room-nose-room-or-looking-space/`

Glossary

This glossary contains some of the common terms used throughout the book and in video editing with their meanings, in alphabetical order.

Artificial intelligence (AI)

A general term used to describe a computer that makes its own decisions based on data it has received.

Archive

The process and name for video and audio files that are stored when no longer used.

B-roll

Video footage that is captured in addition to the main subject or story. B-roll footage is usually of cutaways, cut-ins, or reaction shots.

Bins

Folders that are used to store video and audio clips in your video editing software.

Cache

A temporary file storage location that is hidden on your computer and allows your computer to access files very quickly.

Caching

The process of your computer writing files to your cache.

Clip

A section of video or audio.

Code

A set of written instructions that tell a computer what to do.

Color cast

Where a white, gray, or black image is not pure grayscale but has a hint of color to it.

Color grading

The process of correcting and enhancing the color of a video.

Color management

The act of managing the color of the video footage from camera to final screening, to keep the color consistent with how it was originally shot.

Color model

A theoretical model to describe how colors are mixed to create other colors.

Color space

A 3D space that contains all the color and gamma values available for a particular video standard (e.g., Rec 709).

Color temperature

The color cast of white light on a sliding scale between orange and blue.

Colorist

The person who performs color grading on a video.

Composition

The arrangement of subjects within the video frame.

Continuity editing

A style of video editing where footage that has been shot on different days and times is edited together to look like it all occurred at the same time.

Cut

The abrupt move from one video clip to another without a transition. This can also be used to describe the act of splitting a video clip into two.

Cutaway

A clip that shows other subjects away from the main action of the video (e.g., a shot of the weather).

Cut-in

A clip that shows a more detailed view of the main action (e.g., a shot of the main character's hands).

Data

Information that is stored by a computer.

Database

A place on the computer that stores data. Resolve uses a database to store all the project files.

Digital audio workstation (DAW)

A software application that allows you to edit audio, including music and other recorded sound.

Exposure

The amount of light that enters the camera to create the video's image.

An over-exposed image has too much light, making it look bright. An under-exposed image has too little light, making the image look too dark.

Exposure controls

The settings on the camera that allow you to control how much light is recorded to create the video's image.

Frame rate/FPS

The speed at which the frames of a video image are played back to the viewer. The speed is measured by the number of frames that can be played back in one second. This is described as **frames per second (FPS)**.

Frames

A video image is made up of a series of photographs that are shown back rapidly to give the illusion of movement (see the *Frame rate/FPS* section). Each of these photographs is called a frame.

Framing

The act of moving the camera in relation to the subject to change the composition of the image.

Gamma

How an image displays the contrast between the brightest and darkest part of an image. This is based on a curve that shows the relationship between the original video level and its perceived brightness on a screen.

Graphics

A picture that is created by a computer, such as a logo or animation.

Greenscreen

A term commonly used to describe the process of shooting a video with a green background that is later removed by a computer and replaced with another still or moving image as a background.

Gray card

A card that is neutral gray with no color cast that is used to help white-balance the camera.

Hard disk drive (HDD)

Also called a hard drive. A hard drive stores data that your computer uses, such as files and applications.

ISO

Short for **International Standards Organization**, it is a camera setting that helps control the level of exposure of an image.

J-cut

A video cut where the sound is heard before we see the video image.

L-cut

A video cut where the sound is heard after the video image is no longer seen.

Lead room

A video composition technique where more room is given in front of a subject to allow movement of the subject into the frame (e.g., a car driving will be framed to have more space in front of it than behind it).

Looking room

The same as lead room, but usually used to describe having more space in front of a person than behind so that they appear to be looking into the frame rather than out of it.

Media

A general term for any files used in your video edit, including video, audio, graphics, and still pictures.

Neural Engine

Blackmagic Design's name for the AI function in DaVinci Resolve.

Neutral

An image or grayscale with no color cast to it.

Non-linear editor (NLE)

The technical name for video editing software, as opposed to old-fashioned film or tape-to-tape editing. It is named *non-linear* as you place the clips in any order on the **Timeline**. In old-fashioned linear editing, you would have to start at the beginning of the film and add each piece of footage one after another, unable to insert an edit without it overwriting the rest of the film. In an NLE, you can start at any point in the film and move all the clips around at any time.

Plugins

Small programs that add new abilities to the larger host program.

Project

The name of the file that contains all the video editing decisions of an NLE.

Reaction shot

A video of someone visually reacting to the main action of the story (e.g., an interviewer nodding in response to an interviewee's answer).

Render

The act of a computer processing data into a final viewable image.

Resolve

An abbreviation of DaVinci Resolve.

Run and gun

A style of filming often used for documentary and reality TV, where the video is captured quickly with no time to preplan the shoot. As the camera is handheld to help capture the scene quickly, it can lead to shaky footage.

Solid state drive (SSD)

A hard drive that has no moving parts and is more durable, lighter, and faster than regular hard drives.

Sound effects (SFX)

Short sound clips recorded to add sounds that are not in the original video to emphasize the action in the video (e.g., a door hinge squeaking).

Split edit

A cut where the sound is separated from the video, creating either a J-or L-cut.

Stills

Another word for photographs. Short for still images (compared to moving images or "movies").

Thumbnail

A small picture that represents the content of the video file or clip.

Three-point edit

An editing technique that uses an **In** and **Out** point on either the **Timeline** or the source clip and a third point marked by a playhead or either an In or Out point to determine how much of the original clip is to be used and where to place it on the **Timeline**.

Timeline

A **Timeline** is a line that visually shows events in the order that they happened. In editing software, the **Timeline** is where clips are placed and arranged in an order that represents the timing of the story being told.

Tools

Features that give you specific functionality within software by clicking an on-screen icon or button without needing to apply a plugin (e.g., the **Insert Clip** tool).

Transitions

Video effects that slowly transition from one video clip to another.

Visual effects (VFX)

Computer-generated effects that change the look of a video image.

White balance

The act of changing the color of the image to compensate for a color cast to make the white and black areas of the image neutral (e.g., adding more orange to compensate for a blue-tinted image).

Answers to Questions

Here are the answers to the questions that were at the end of the chapters in this book. Use these answers to check your understanding of each chapter.

Chapter 1, Getting Started with Resolve – Publishing Your First Cut

1. False. Adding new clips on **Track 1** ripples the rest of the clips on the **Timeline** to make room for the new clip, whereas adding a new clip-on Track 2 can overwrite any existing clips on **Track 2** without affecting the clips on Track 1.

2. YouTube, Vimeo, and Twitter.

3. **Smart Insert**, **Append to End**, and **Ripple Overwrite** work on **Track 1** only.

4. **Close Up**, **Place on Top**, and **Source Overwrite** all place clips on **Track 2**.

5. False. *Timecode* is the term used to describe the position of the playhead on the **Timeline**. It is displayed in hours, minutes, seconds, and frames (HH:MM:SS:FF).

Chapter 2, Adding Titles and Motion Graphics

1. False. A .DRP file does not contain any media. The media will need to be imported separately and relinked to the project within Resolve.

2. False. Both **Text** and **Text+** titles offer you all the fonts available on your computer.

3. E. Cross Dissolve

4. True.

Chapter 3, Polishing the Camera Audio – Getting It in Sync

1. False. The best microphone to use is the one closest to the subject.

2. False. You can also sync separate audio to the video using timecode and by manually syncing using a visual reference in the video.

3. True. On the **Cut** page, navigate to **Effects** > **Audio** > **Fairlight FX**.

4. True. **Voice Isolation** can be found in the **Audio** section of the **Inspector**.

Chapter 4, Adding Narration, Voice Dubbing, and Subtitles

1. True

2. False. Resolve supports SRT and WebVTT for web-based subtitles as well as the broadcast-friendly IMSC1 or DFXP subtitle formats.

3. False. **Fairlight** is not a plug-in; it is a fully working DAW accessible right within DaVinci Resolve.

4. Pressing the *M* key in **Fairlight** will add a **Marker** on the **Timeline**.

5. True.

6. False. It stands for **Automated Dialogue Replacement**.

Chapter 5, Creating Additional Sound

1. True.

2. False. On the **Cut** page, navigate to **Effects** > **Audio** and choose either **AU Effects** or **VST Effects**.

3. True.

4. False. You can import SFX into any **Project Library** (database).

5. True.

Chapter 6, Working with Archive Footage

1. False. **Fusion FX** are also available in the free version of DaVinci Resolve.

2. False. You can also change the speed of clips on the **Cut** and **Edit** pages.

3. True.

4. True.

Chapter 7, Stabilizing Shaky Footage

1. False. Capturing stable footage is usually the better option as it saves time in post-production, and it will not lose resolution if stabilized in post-production.

2. False. You can either use the **Stabilize** tool in the clip's **Toolbar**, or the **Stabilize** controls in the **Inspector**.

3. False. The stabilization controls in the **Inspector** are the same on both the **Cut** and **Edit** pages.

4. True. There are the stabilization controls in the **Tracker** palette on the **Color** page, but you also have the option to use the **Classic Stabilizer**. Both allow you to select how you want your footage to be tracked and then stabilized.

Chapter 8, Hiding the Cut: Making our Edit Invisible

1. False. A Cutaway or Cut-in is a type of shot whereas a **Split Edit** is a type of transition from one clip to another.

2. False. Continuity editing is where we edit the clips on the **Timeline** so that they appear to be continuous within the same timeframe, even though they may have been filmed in a different order on different days.

3. True.

4. False. A Cutaway is a B-roll clip used to illustrate the main action of the scene.

5. True. It can be found with all the other transitions in the **Effects** panel.

6. True.

Chapter 9, Adding Special Effects

1. False. It is key to light our background evenly and with no shadows, as it is easier to remove a single color in post-production without affecting the subject.

2. False. The **3D Keyer** has nothing to do with 3D images. It can be used like any other keyer to remove a colored background, regardless of whether the video is 2D or 3D.

3. True.

4. False. There is also an option to remove red background spill.

Chapter 10, Split Screens and Picture-in-Picture

1. False. There are several ways to create a split-screen effect on the **Cut** page.

2. False. We can use **Decompose in Place** to access the clips again.

3. False. It needs to be on the topmost video layer.

4. True.

Chapter 11, Enhancing Color for Mood or Style

1. False. The human eye has light-receiving cones that are more sensitive to red, green, and blue wavelengths of light.

2. True.

3. False. You can use the **Detail Size**, **Low Threshold**, and **High Threshold** sliders to select which parts of the image you want to change the contrast for.

4. False. There is an **Auto Color** tool in the **Tools** bar on the **Cut** page.

5. True.

Chapter 12, Studio-Only Techniques – so much quicker

1. False. It is found in the **Inspector**: **Inspector** > **Video** > **Transform** > **Smart Reframe**.

2. True. But only if the clip has metadata that matches the search terms that the **Smart Bin** is looking for.

3. True.

Index

`Packtpub.com`

Subscribe to our online digital library for full access to over 7,000 books and videos, as well as industry leading tools to help you plan your personal development and advance your career. For more information, please visit our website.

Why subscribe?

- Spend less time learning and more time coding with practical eBooks and Videos from over 4,000 industry professionals
- Improve your learning with Skill Plans built especially for you
- Get a free eBook or video every month
- Fully searchable for easy access to vital information
- Copy and paste, print, and bookmark content

Did you know that Packt offers eBook versions of every book published, with PDF and ePub files available? You can upgrade to the eBook version at `packtpub.com` and as a print book customer, you are entitled to a discount on the eBook copy. Get in touch with us at `customercare@packtpub.com` for more details.

At `www.packtpub.com`, you can also read a collection of free technical articles, sign up for a range of free newsletters, and receive exclusive discounts and offers on Packt books and eBooks.

Other Books You May Enjoy

If you enjoyed this book, you may be interested in these other books by Packt:

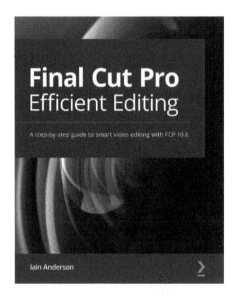

Final Cut Pro Efficient Editing

Iain Anderson

ISBN: 9781839213243

- Understand the media import process and delve into media management.
- Effectively organize your footage so you can find the right shot quickly.
- Discover how to assemble a rough cut edit.
- Explore trimming and advanced editing techniques to finesse and finalize the edit.
- Enhance an edit with color correction, effects, transitions, titles, captions, and much more.
- Sweeten the audio by controlling volume, using compression, and adding effects.

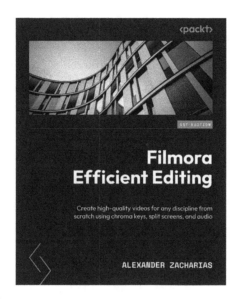

Filmora Efficient Editing

Alexander Zacharias

ISBN: 9781801814201

- Navigate Filmora's interface with ease.

- Add and manipulate audio using audio tracks.

- Create high-quality professional videos with advanced features in Filmora.

- Use split screens and Chroma keys to create movie magic.

- Create a gaming video and add humor to it.

- Understand career prospects in the world of video editing.

Packt is searching for authors like you

If you're interested in becoming an author for Packt, please visit `authors.packtpub.com` and apply today. We have worked with thousands of developers and tech professionals, just like you, to help them share their insight with the global tech community. You can make a general application, apply for a specific hot topic that we are recruiting an author for, or submit your own idea.

Hi!

I am Lance Phillips, author of *Video Editing Made Easy with DaVinci Resolve 18*. I really hope you enjoyed reading this book and found it useful for increasing your productivity and efficiency.

It would really help me (and other potential readers!) if you could leave a review on Amazon sharing your thoughts on this book.

Go to the link below or scan the QR code to leave your review:

`https://packt.link/r/1801075255`

Your review will help us to understand what's worked well in this book, and what could be improved upon for future editions, so it really is appreciated.

Best wishes,

Lance Phillips

Download a free PDF copy of this book

Thanks for purchasing this book!

Do you like to read on the go but are unable to carry your print books everywhere?

Is your eBook purchase not compatible with the device of your choice?

Don't worry, now with every Packt book you get a DRM-free PDF version of that book at no cost.

Read anywhere, any place, on any device. Search, copy, and paste code from your favorite technical books directly into your application.

The perks don't stop there, you can get exclusive access to discounts, newsletters, and great free content in your inbox daily

Follow these simple steps to get the benefits:

1. Scan the QR code or visit the link below

https://packt.link/free-ebook/9781801075251

2. Submit your proof of purchase
3. That's it! We'll send your free PDF and other benefits to your email directly

www.ingramcontent.com/pod-product-compliance
Lightning Source LLC
Chambersburg PA
CBHW062058050326
40690CB00016B/3132